'This book w...
the way you think. Forever.'
COSMOPOLITAN

'An inspirational memoir from an extraordinary
woman. *The Heat of the Moment* does for the fire
service what Adam Kay's *This Is Going to Hurt*
did for the NHS. A humbling, jaw-dropping read
which redefines what is possible in life.'
VIV GROSKOP

'An immersive insight into a job which few of
us could do, by a woman whose award-winning
research into decision-making in the emergency
services has transformed policy at a global level.'
STYLIST

'A vivid reminder of the horrors that
firefighters face daily – and the debt
of gratitude society owes them . . .
This book challenges assumptions about who
firefighters are, and about what women can do.'
GUARDIAN

'Draws on stories from the frontline of
firefighting to show us what it means to be
human in the face of disaster.'
SUNDAY POST

The Heat of the Moment

Moment

*A Firefighter's
Stories of Life and
Death Decisions*

DR SABRINA
COHEN-HATTON

TRANSWORLD PUBLISHERS
61–63 Uxbridge Road, London W5 5SA
www.penguin.co.uk

Transworld is part of the Penguin Random House group of companies
whose addresses can be found at global.penguinrandomhouse.com

First published in Great Britain in 2019 by Doubleday
an imprint of Transworld Publishers
Black Swan edition published 2020

Epigraph on p. 1 extracted from 'This Week in Fiction: Mohsin Hamid' by
Cressida Leyshon, *The New Yorker*, 16 September 2012. Epigraph on p. 23 extracted
from *Managing* by Harold Geneen, reprinted by permission of HarperCollins
Publishers Ltd © Harold S. Geneen 1985. Epigraph on p. 107 extracted from *The
Metaphoric Mind* by Bob Samples with kind permission of Cheryl Charles. Table
reproduced on p. 189 taken from the Joint Emergency Services Interoperability
Principles Joint Doctrine: www.jesip.org.uk. Table on p. 232 reproduced with
kind permission of Phil Butler, Rob Honey and Sabrina Cohen-Hatton.

A CIP catalogue record for this book
is available from the British Library.

ISBN 9781784163884

Typeset in 10.92/13.7pt Minion Pro by Jouve (UK), Milton Keynes.
Printed and bound in Great Britain by Clays Ltd, Elcograf S.p.A.

Penguin Random House is committed to a sustainable
future for our business, our readers and our planet. This book
is made from Forest Stewardship Council® certified paper.

MIX
Paper from
responsible sources
FSC® C018179

1 3 5 7 9 10 8 6 4 2

To Gabriella
Little and fierce
I will raise you like you breathe fire
Never give up.

To Mike
For your enduring patience and bottomless cups of tea.

I love you both
Always.

Contents

Author's Note

Throughout this book I primarily explore the perspective of an incident commander. However, no large or complicated incident can be commanded alone. There is always a group of additional officers. They form a command team and take responsibility for certain elements of the incident. Some are in charge of the activity in a specific area or sector; others are in charge of functions, such as logistics or crew welfare. The roles aren't predetermined – they change at every incident – and so too do the people available to fill them.

Some additional material to clarify some of the fire service terms, including our ranks, different responsibilities and equipment, has been added to the back of this book (see p. 263). Please do check it out if you're ever in doubt.

Foreword

Thick, black smoke spills from the roof of a house on the other side of the road; bursting from between the tiles. Flames lick at one side of the building, a medley of yellow and orange, and on the other side a man is leaning out of a window, waving his arms and shouting.

You know what to do. You know who to call. You know that help will come. You trust that when you need it – for your friends, or your family, or even for yourself – someone will come to rescue you.

My father always said that it takes a special kind of person to rush in when everyone else rushes out. I have worked as a firefighter, and more recently as a commander, for nearly twenty years, and I know that he was right. Firefighters are brave enough to face flames; to endure dark, dense smoke; to defy natural instinct and trust that there will be a way out. We are compassionate. We hate to watch people suffer, to witness the injustice of a life snatched away prematurely. We will run through fire to prevent that from happening and to save the life of someone we've never met. We are dedicated. We all want to quit sometimes. Yet we carry on. Because firefighting is such a huge part of our identity that it sometimes feels impossible to disentangle ourselves as people from ourselves as firefighters.

Behind every rescue is a rescuer; behind every rescuer is a team of people, making life-determining decisions and working to save others. Within that team, commanders such as myself are also in the fortunate position of being trusted by the firefighters to pull together the plan, making sure the right people are in the right place, at the right time, to save lives.

We love our work. It inspires and challenges us; it encourages us to be better people, to get fitter, to fight harder. We spend our days at devastating, life-changing incidents, and we are incredibly privileged to have vulnerable people put their faith in us in their darkest hours. We want to do our best work every day. We *need* to do our best work every day.

The perception of firefighters is ever changing – sometimes we are hailed as heroes, and sometimes people think we fall short of the mark – but, for most of my career, firefighters have been held in very high esteem. You only have to dial three digits and, like superheroes, we will appear exactly where we're needed.

The reality, of course, is that firefighters are human beings. Our instincts drive us to push through the most unforgiving of environments but we have limitations. We have the same fears and flaws as anyone we rescue, but we rely on our skills, confidence and a certain amount of luck. Sometimes it can be too much and we suffer for it. We are just ordinary people, working together to achieve extraordinary things.

It might be the superhero myth that also contributes to the idea of a typical firefighter. Are you imagining a good-looking man; tall, dark and handsome, with a brooding stare? You won't be alone. Contrary to popular belief our nation's fire stations aren't populated by underwear models, but whatever comes to mind when you think of a firefighter, I very much doubt that you're envisaging some-one who looks like me.

I'm five foot one and tip the scales at eight stone. I have long dark hair and like a manicure. I'm a mum. I have a PhD in neuroscience. I'm a nerd. I am not most people's idea of a traditional firefighter but I became one at eighteen, and I've been working my way steadily through the ranks ever since. I now serve as a deputy assistant commissioner in one of the largest fire and rescue services in the world.

I wish there were more women in the fire and rescue service because, at the moment, I'm a bit of an anomaly.* As a result, I've spent much of my career pushing back against stereotypes. I've lost count of the number of well-meaning but ill-informed people who blink at me in disbelief when I tell them what I do. I used to feel frus-trated and irritated by their reaction. However, in a way, the freedom to be different – really different – has been incredibly empowering. I've had the opportunity to define my own boundaries.

Early in my career, I was called to attend what turned out to be a pivotal incident. I rode there in the back of a fire engine quite sure that someone I loved had been badly injured at the fire. It was the most miserable, terrifying journey of my life, and it gave me an insight into how our roles sit on the fringes of real people's lives and real tra-gedy. A firefighter had been injured and, while I was

---

* At present, there are more people named Chris in chief fire officer roles than there are women chief fire officers. Women make up 5 per cent of the number of firefighters in the UK, but that statistic is improving. When I joined, it was just 1 per cent.

relieved that it wasn't my fiancé, I felt the most overwhelming sense of guilt.

The situation had been misinterpreted, which resulted in a serious injury to one of our own. As I later found out, he wasn't the only firefighter to have been injured in the course of his duty. In fact, thousands of firefighters are injured at work each year and most of those accidents happen as a result of human error. These people are some of my best friends, and while the mistakes often aren't their fault, they have the potential to cost them their lives. I found myself asking questions of my colleagues, of the service, and of myself. I needed to know how we could give firefighters the best chance of saving lives without risking their own.

I was determined to make a meaningful difference, and I found that the best way to reduce human error is by examining the way we respond during emergency incidents. So I studied psychology while I was a firefighter, just a few years into my service. I went on to research human instincts, our natural biases, the way that our brains function under stress. After I completed a PhD in Psychology – in behavioural neuroscience specifically – I worked alongside colleagues in the fire and rescue service and at Cardiff University to apply my findings to real-world situations. I wanted to be more effective as a firefighter – a better decision-maker, a better leader and a safer pair of boots on scene. And I wanted that for my colleagues too.

Over the last few years, my work has changed the way we operate on the front line and, in particular, the way we train. I've worked with firefighters all over the world, sharing our findings, and I'm delighted to now have the opportunity to share them with you. These are lessons that have benefited us in the fire and rescue service, but they are just as applicable to everyday life as they are to situations of life or death. Can you trust your gut? What should you decide when neither outcome is what you want? How can you work with others to make the best decisions collectively?

The Heat of the Moment is split into ten chapters. Each explores a different element of decision-making and, I hope, offers an insight into my world – from the scenes of devastation and crisis, through triumphs of bravery, to the quieter moments when these assumed heroes and heroines question themselves.

I will reference a great many real-life incidents and some training exercises throughout this book. Some of them are already a matter of public record, but I have also included a number of lesser-known incidents that, while important to me personally, are not widely recognizable. For some, I have changed everybody's names – from firefighters to other emergency services personnel to the casualties – as well as their physical descriptions, sometimes their roles and the location of the incident. I have also, at times, amended or avoided mention of a particularly distinctive detail (a specific model of car, for example, or an unusual aspect of the response). Indeed, in a few instances, I have blended together my experiences from several incidents or exercises and presented them as one, and on the odd occasion offered my experiences at incidents as exercises and vice versa.

These changes are primarily to preserve anonymity, particularly in cases where there have been fatalities, but they have also been made because this really isn't a book about specific incidents. This is a book about what it means to be human in the face of disaster: the way your brain works, the way you think, the impact of trauma. The incidents simply illustrate the human story.

I hope this book will show you the human side of firefighting, not just its super-human façade. I hope it will change your mind about what a firefighter might look like and encourage you to think instead about the requisite skills and mind-set. I hope too that it will inspire you to imagine yourself in those boots. When you make a decision in the heat of the moment, what will you choose?

1

Trading Places

'Empathy is about finding echoes of another person in yourself.'

Mohsin Hamid

ALL FIREFIGHTERS remember an incident when they thought they were about to come to harm while desperately trying to protect someone. Although we can never promise a way out, if one exists we are often willing to risk our own safety to find it. We are rescuers. We know that we are sometimes the only thing standing between a dying breath and another day.

For me, that incident was a house fire near Cardiff, fifteen years ago. I was in my fourth year of service. It's a memory burned so vividly in my mind that when I recall it I can practically taste the soot.

I wake with a start to the station bells screaming and the dormitory lights flicking on. I sit upright and glance at the clock on the wall: 1.52 a.m. My arms prickle with goose bumps as I throw off the covers and pull on my trousers. It's a warm summer night but the industrial air-conditioning system has been set to arctic and it's freezing. I'm in a mixed-sex dorm, each bed separated by a screen. Most of the watch is out on another call. The remaining four of us rush to the pole, jostling for position. The mood is one of excitement, full of anticipation. I'm small so I slip to the front of the queue and slide down first.

'Too slow, old men,' I call.

There are twelve firefighters on my watch and they're all like brothers to me. I hit the floor with a thud and tear the message from the tele-printer.

'House fire,' I shout. 'Persons reported!'

The atmosphere sharpens. We know that someone needs help. I hand the printer ticket to Bert, the watch manager. He's a seasoned veteran and remains one of my biggest inspirations. Even now, I often look at a raging fire and think, 'What would Bert do?'

He scans the details. 'Everyone on the truck – quickly,' he barks.

Within seconds, we're all in position. The engine is revving, filling the appliance bay with acrid exhaust fumes. The fire engine doors slam shut in quick succession and we pull out of the station. The sirens scream and the cars stand still as we plough across the road and through the junction.

I'm assigned to the breathing apparatus team with another firefighter, Jack. We are to go into the residence first and search for missing people. We get rigged, unclipping our seat belts and putting on our breathing apparatus 'sets'.* Strictly speaking, we're not meant to do this while we're still driving, but we're not willing to sacrifice those extra seconds – seconds that could be the difference between life and death. I reach behind and turn on my set; the unmistakeable hiss of air entering the hose tells me it's ready for use. This is my lifeline.

I look over at Jack.

'Ready?' I ask. I pull my straps tightly.

'Yep,' he replies. 'All good.'

'Don't suck that air too quickly, will you, big man?' I tease. 'I don't want you holding me back with a cylinder change when I'm still three-quarters full.'

'Don't make me do all the work and I won't need to,' he quips with a broad grin and a wink.

---

* These are harnesses that are rigged up to a cylinder of compressed air and a mask. They supply fresh air so that firefighters can breathe when we enter a fire. For more information see Notes, p. 265.

'Fat chance of that,' I say.

Pete – an experienced firefighter on our watch – is driving. Fast. I wind down my window slightly. Sometimes you can smell the fire before you see it. I look outside, scanning the night for signs of the blaze. I can feel my heart beating a little faster than I wish it would. We might all look calm on the outside, but on the inside we're anxious – excited even – and eager to help. We know this will be physically demanding – all serious house fires are – but the sense of anticipation is fuelling us.

Bert looks over his shoulder.

'Right,' he says. 'We might have people inside so I want you straight in when we pull up. The second crew are on their way, but we'll be on our own for a while. Pete will be operating the pump as well as entry control, so make sure you have the same radio channel. And watch yourself,' Bert continues. 'I've been to this street before and the place was full of drugs. Watch out for needles. And if you find anything weird, like a machete, will you please leave it where it fucking well is this time, Sab.'

'Point taken, boss,' I say. 'But it was a big machete.'

'She's right, Bert. It was a big machete,' Jack adds.

'Over there,' I say. 'Smoke on my right.'

Thick clouds billow over the ridges of the rooftops. The smoke glows a dirty shade of grey against the stark black sky, reflecting the angry blaze beneath. The rows of terraced houses sparkle like tinderbox decorations. I wonder who might be inside the bedrooms, sleeping, completely oblivious. These houses tend to have a common roof void where the attic spaces are connected and fire can spread very easily and very quickly. I shuffle impatiently in my seat.

We spin around a corner and the fire engine screeches to a halt. We've all been to this area before. Poverty is the norm and social problems are rife. Crime is high and hope is low. It's a dramatic contrast to the newly regenerated areas of the city, full of designer shops and young professionals in expensive new apartments. Less than a mile apart but the gulf between them couldn't be wider.

We've stopped a few doors down from the flaming house. If we park too close, we risk the fire engine being destroyed in a backdraft. The appliance is a lifeline for the firefighters inside. Without it, we wouldn't stand a chance.

A small crowd has gathered. The police haven't arrived yet so we'll be dealing with onlookers as well as with the fire.

We leap off the truck. I rush to the rear locker and throw it open. I grab the hose-reel, turn, and run a few metres, dragging it off the barrel and laying it in a tight zigzag so that we can easily haul it through the door without getting stuck.

Jack and I pull on our masks and helmets. We check each other for gaps where the fire might burn us. I tug his sleeve over his glove to cover the exposed skin on his wrist. We give each other the universal thumbs-up and we're off. Jack grabs the branch* and I pull the hose behind us.

The front door is open and we're in. It's pitch black and the thick smoke is disorientating. I keep my left hand on the wall. It's warm, but I think the fire is above us. Jack pulses water into the ceiling to cool the hot gases and I drag as much of the heavy hose into the house as I can – enough to reach the furthest point, and then a bit more. It's a large terrace, I'd

---

* The nozzle on the end of the hose-reel used to control the flow of the water.

guess five by fifteen metres, so I pull in about thirty metres of hose using the couplings to help estimate the length.

We push through the corridor. It's dark and hot. I feel the ripple of old flock wallpaper under my hand. I wonder who lives here. Perhaps it's an elderly person. If so, their mobility might be an issue. I shuffle along and feel my foot scrape against something. We're near the stairs; it feels like shoes. I reach forward with my right hand. I grip something cylindrical. It's solid and narrow. A walking stick? No. A handlebar. A pram. Shit, it's a pram.

I feel a surge of urgency knowing there could be a child in here. A child that might not be able to crawl, or even cry for help. I shout to alert Jack. He has our radio comms – our only connection with the outside world. He relays to Bert that there's a pram, that there might be a child – or children.

'Upstairs,' I shout to Jack. 'If there are kids, they'll be upstairs.'

We stamp on each step before shifting our weight on to it. If it will take a stamp, it hasn't yet burned through. The temperature increases significantly as we ascend. Hot air rises, after all. We're getting closer to the fire. We reach the top and the heat begins to restrict my breathing, building in my chest. It's like the moment you step into a sauna and it's too hot to draw breath. But the sensation doesn't ease. You don't get used to it. It just gets hotter and hotter.

I can feel the heat of the metal buckles of my set through my fire kit. If they stay in one place for too long they'll burn me, and I could do without that happening again. I keep my left hand on the wall. The old plaster disintegrates as I run my fingers along it. I reach a door frame and step inside.

The room is full of smoke. I have no idea what it is – a

bedroom? A bathroom? – but it doesn't matter. People can be found anywhere in a fire. Jack and I shuffle around, feeling everything, trying to paint a picture through touch. My hand falls on something solid and round. It's a sink. I accidentally nudge open a tap and water gushes out, drenching my glove.

'A tap,' I call. 'It's a bathroom.'

'Careful,' shouts Jack. In such intense heat, the water can turn to steam. It's like holding your hand above a boiling kettle. The scalds take weeks to heal. Receiving a burn in a fire – however minor – complicates the situation. You'd be expected to leave. However, no one ever does. We do everything we can to finish the job. Sometimes you don't even notice until you get out and the adrenaline drains away.

'I hope there aren't needles,' says Jack. 'It gives me the creeps when Bert says something like that.'

'I know, but be careful where you put your hands,' I reply.

The bathroom is empty. We move on. The fire is in the room opposite. We think it's a bedroom. The door has burned through and the flames lick the frame. They are roaring, vibrating over the hiss of my breathing apparatus. I peer inside. Tongues of red and orange are wrapped around the curtains and its rail. Pieces of fabric are dripping to the floor. No one in there will have survived. Jack puts a message back to Bert to find out how long our back-up crew will be. They are coming from the other side of the city, so a while.

'I'll keep the fire in check,' Jack yells. 'You search the other room.'

The fire hasn't jumped the corridor, so there could still be someone alive inside.

I hesitate. Strictly speaking, we're meant to stay together. But we need to stop the fire escalating, so at least one of us has to be putting water on it and trying to prevent the spread. If we both do that, then no one is searching for the people – possibly children – who we know are relying on us.

Jack is crouching, pulsing bursts of water towards the ceiling to cool the hot gases. He's got this. I push on.

I go to the next room. I map my movements, noting every step I make. I need to know I can find my way back should Jack need me. I crawl into the room on my knees. I try to stand but it's too hot. Knowing time is short, I edge quickly along the perimeter and feel a wardrobe. I've found children hiding in wardrobes before, but not today. It's empty. Just a few toys. I move on. Beads of sweat trickle down my face, soaking the rubber inside my mask. It will be worse for Jack – he's much closer to the fire.

I call out to him. He's fine, so I move on again.

I continue to search the room, keeping my left hand on the wall or on a piece of furniture, mentally plotting where I am and where I've been. Touch is my only useful sense. It's too dark to see. All I can hear is the roar of flames and the hiss of compressed air. All I can smell is the rubber of my mask.

My hand hits something hard and metal clinks against my buckles. A bed? It has legs, but they're too narrow to be a bed. I reach up, translating the solid lines into an image. Narrow vertical bars. A mattress.

It's not a bed, but a cot. I'm not looking for a child; I'm looking for a baby. I feel beneath the cot. Nothing. I try to stand up but it's too hot at head height. The sting of heat prickles at my ears through my protective hood, like hot little needles being held against my skin. I grimace and push through. If I'm struggling, even with all of this protective gear, then I can't imagine how a baby might be suffering.

I stand quickly and reach inside the cot. I manage a few seconds and then kneel back down to take a break from the heat. I couldn't feel anything but I'm still not convinced. The baby must be here somewhere.

I leap to my feet and do a thorough sweep of the whole cot. A pram, a cot, toys – there *must* be a child. But there isn't. At moments like this, you doubt your rational self. It's like when you've lost your keys and keep returning to the place you *think* they are, even though you've checked that spot a dozen times already.

There's nowhere left to look. A toddler might have climbed out of the cot.

Or a baby could be with its parents in their bed.

I head to another room.

Something doesn't feel right. I can't put my finger on it. It's a gut instinct, and often gut instincts are based on experiences lodged in your subconscious; you just don't know it at the time. My skin stings and my eyes widen as my senses sharpen, alert to any clues as to what it is that's making me feel so uneasy. I can hear a rumbling noise; a muffled, throaty sound. I come to some tall wardrobes; I yank the doors open and rummage through the clothes. They are piled in and it's difficult to feel behind them for any sign of a person. Nothing. I carry on. I can still hear that noise.

'Sab, you're gonna have to hurry,' Jack shouts. 'This is getting too hot.'

'I'm not done,' I reply.

'We don't have long,' he says.

I push on, the noise persisting. I find a bed and nudge

my boots underneath it as far as I can in case someone is hiding there. I don't feel anything. I reach up to the mattress and pull off the covers and pillows. Still nothing. But that noise. It's all I can focus on.

'Sab – enough, mate. We gotta split.'

'Just two seconds, Jack. Something's not right.'

I reach over the bed and come into contact with a wall. I couldn't feel it when I checked beneath with my feet – my legs were too short. That means there must be more space beneath to search. The sound is getting louder. I pull the bed away from the wall. It's heavy and it drags on the carpet. I crawl on to the mattress and reach down the other side.

I feel something. No, not something. Someone. An adult. They must have been wedged between the bed frame and the wall.

'Jack, I've found someone,' I shout. I drag the person around the bed frame. I can't yet tell if it's a man or a woman, but they are big and they are heavy, like a dead weight. I'm wearing 15 kilograms of breathing apparatus and around 8 kilograms of fire gear, so technique will be everything if I'm going to move them. I slide my arms underneath theirs and lock my hands in a fist on top of their chest.*

'It's all right,' I say. 'We've got you. Where's the child?'

The person's only response is that same throaty gurgle. It's animalistic. I'm afraid they might be choking. We have

---

* The shoulder lift you see in the movies is an historical technique and not something we use in practice any more. We work in teams when carrying a person and try to keep their head low to the ground where there is more air.

no time. I rush back around the bed, searching urgently for the child. This person could have been holding a baby, keeping their child tucked beneath a protective arm. But there's no one here.

Jack comes in to help with the casualty.

'There's still no sign of the kid,' I say, 'but we need to get this one out. How are you for air?'

I want to keep searching, but I'm afraid this person is dying. The noise they are making could be a 'death rat-tle' – normally caused by a build-up of saliva in the throat and often a sign that someone is near death – but I've only heard it a few times before.

'I'm down to seventy bars,' Jack replies. 'You?'

'One-twenty,' I say. Seventy bars of air isn't much. He's used two-thirds of his cylinder already. I have more left because my lungs are smaller. My air goes much further, and I'm keen that Jack – my friend as well as my colleague – doesn't run out too soon. In this job we share the burden.

'I'll take the heavy end,' I say. 'You get the legs.'

We lift the casualty, descend the stairs as quickly as we can, and bring them out to the waiting paramedics.

We emerge to bright street lights that make me wince while my eyes adjust. I look down. Our casualty is a man with a thick mop of tangled brown hair. He's young – around thirty. I realize, horrified, that the throaty rumbling – the same noise that tore me away from my desperate search to find a child – is this man's snoring. I peel off my mask.

'Sir,' I shout at him. 'The child? Where is the child?'

Nothing. I'm furious.

'Sir!'

The paramedic looks at me. 'You'll be lucky,' she says.

'We've just pulled an empty bottle of methadone from his pocket. You've got no chance of a reply.'

We haven't finished the search. I still have half of that room left to do and the entire third bedroom. And we haven't searched downstairs. There might still be a child in there.

This is my moment. The moment when I would have sacrificed everything – my own life – to save that child.

I cannot promise a way out – I know that – but I can promise to do everything I can to find one. That is the vow I make when I walk into a fire. On that day, I didn't keep my word. Standing there, beneath the harsh street lights, in the warm summer air, I felt that I'd failed at my job.

I was angry that this guy had got so out of his head that he had slept through a fire that might have killed his kid. I was angry, but then I was also so sad for him. I wondered what he would think when he learned that we had saved him but had failed to rescue his child. How he would feel to wake and discover that he'd never hold that little body again? Or read, or talk, or sing to his daughter or son? Or hear them giggle? He wouldn't see them grow up. He wouldn't witness their first day at school. He wouldn't attend their wedding. He'd have none of that. Regardless of how much his current state of intoxication might have contributed to things, I felt as if I was the one who had failed. It wasn't a rational response. It was emotional.

Most parents would sacrifice themselves to save their child. In that moment, I'd have sacrificed myself for his too.

'Get fresh cylinders,' I yell at Jack. 'We need to get back in there. We are *not* done.'

Bert rushes over. 'The second pump is here. Stand down. I'm putting them in.'

'But we're good, Bert.' I know exactly where to go, exactly what to do. I can see the layout of the rooms so clearly in my mind. 'It needs to be us.'

'You're not going back in.' Bert is resolute. 'You're going to tell the other crew everything you know and then you're going to drink this water, change your cylinders and re-test your sets. Then rest for a few minutes. Depending on their progress you might be in again shortly and I need you ready.'

Bert is right. I drop my set down beside Jack's.

'Stop it,' Jack says to me.

'Stop what?' I frown.

'Stop what you're doing right now. Punishing yourself. We found one. We did good.'

'Jack, there was a pram, a cot, toys. It's a kid's house. There's a kid in there. That guy is out of his head. We don't even know if it's his house. But what I do know is that it's the middle of the night, people are at home asleep in their beds and we haven't found the kid that should be sleeping in that cot.'

'I know, mate, I know.'

I have a sip of water and get ready to go back in. We stand at the entry control point, waiting eagerly for news.

Bert comes over. 'All persons accounted for,' he says.

I freeze.

'The child?' I ask.

'The mother and baby weren't in. There'd been an argument and she was staying with her parents. Brief's changed from search and rescue to firefighting.'

I look up at the sky and exhale. My hunched, tight shoulders drop down and I relax. I blink back tears and swallow hard. I can't explain that sense of relief. The baby that I had searched so desperately for wasn't there.

The conditions inside had been punishing. Even I – with all my protective gear – had found them challenging. It would have been impossible for a child to survive.

I have lost count of the number of complete strangers whose pain has reverberated through me as if it were my own. The wife who lost her husband in a road traffic collision. The child who lost a parent in a fire. The siblings separated for ever by the accident that stole one of their lives. The list is long and that's just my list. All of these people woke up one morning expecting another normal day and instead found their world irreversibly and eternally changed.

This pain – the pain I experience – is empathy. It's what drives me to go that little bit further, push that little bit harder. As well as empathizing with those who rely on us, I also empathize with those I serve alongside and those I now command. It is vital that I understand not only the implications of what I'm asking my colleagues to do, but also how they might feel.

This might mean in a physical sense. I understand the conditions of fire – what it feels like to enter a burning building, carrying 23 kilograms of extra weight, and to work hard. I know the taste of sweat inside my mask. I know the feeling of part of my helmet melting just above my head. But I also understand how being a firefighter can have an emotional impact. I know how it feels to crawl on your belly to reach someone on the other side of the burning room only to put your hand straight through their ribcage and realize they're nothing but charred remains. I know what it feels like to see the faces of loved ones fall when you come out empty-handed. I know how it feels to see the human face of hope ebb away. I know what it feels like to push on when you just want it to be over.

My appreciation of operational empathy was sharpened by one particular fire I attended. That incident would go on to drive one of the most significant pieces of psychological research for the British fire and rescue service and influence further studies into the psychology of incident command.

I was part of a crew called to a scene where a firefighter had been severely burned. The incident involved only one fire engine and I knew that my husband-to-be, Mike, was part of that crew. There was a one-in-four chance that he was the injured firefighter.

I remember hearing the bell ring and rushing into the appliance bay. A few crew members were already at the teleprinter and they were huddled around the slip of paper.

'What have we got?' I asked.

No one replied. I looked at Bill, the crew commander.

'A firefighter injury,' he said eventually. 'Someone's been burned. It sounds bad.'

I pulled on my boots and yanked my braces up over my shoulders. I reached for my jacket and, as I put my arm though the sleeve, Bill continued.

'It's bravo-two-one, Sab,' he said. 'It's Mike's crew.'

I stopped immediately, jacket half on. I heard his words, but I felt completely detached from the moment. It was as if I was in a film, the main character standing in the centre of the shot as the camera pans across the entire room. My breath caught in my chest. It felt as if I'd taken a killer uppercut to the gut and the wind had been knocked right out of me.

Was I about to become one of those people who wake up in the morning to cornflakes and normality only for their entire world to be ripped apart and changed for ever in the space of just one day?

'Right,' I said quietly. 'Let's go, then.' With that I turned, grabbed my kit and we were flying out of the doors.

My heart was beating so fast. I tried to maintain a quiet, expressionless exterior, but on the inside I was burning. And it was the worst sort of fire; the one that turns everything to ash.

There were five of us on duty that day and I sat in the middle at the back of the truck. All I could think about was Mike. Was it him? Was he hurt? Was he alive? Was he in pain? Was he afraid? At the same time, I tried to focus on what we needed to do next. If I had been thinking straight, I would have been asking questions about the fire. I should have been considering the trauma care equipment we had with us and what we might need.

That journey to the incident ground was the longest four minutes and thirty-seven seconds of my life. I was somewhere between sick and numb the entire time.

We pulled up and I craned my neck to see over Bill and out of the window, desperate for a glimpse to confirm or deny my worst fears. Bill was blocking my view, though, and telling me not to look, to stay on the truck. I nodded but as soon as he got off I was right behind him. I reached for the trauma pack and told myself to get a grip.

I made my way towards the crowd of firefighters crouched down around the injured officer. I could see a pair of legs on the ground, patches of dirt adorning the old, battered blue fire kit, and a pair of boots sticking out from the

huddle. They belonged to whoever had been burned, whoever was lying on the floor. I could hear groaning.

'Get the burns pack,' I heard someone shout. The world
came into sharp focus. I had a job to do and my crew were
relying on me. I couldn't afford to become a hindrance by
shutting down. It took every bit of strength I had to maintain my focus.

'I've got it,' I yelled back.

I was about to step into the huddle when I saw Mike
straighten up. He was fine. The scene spun. My head felt
dizzy and I couldn't feel my legs moving. I bit down hard
on my lip to stop my tears.

Mike could see my relief. He acknowledged it with a
simple knowing glance and we set to work. Still reeling, I
opened the trauma pack and took out the burns masks,
trying desperately to steady my shaking hands.

'Sterile water,' I said. 'I need more.'

Mike passed me some sachets and I soaked the masks. I
looked at the man lying in front of me. It was Steve. Red
burns were etched across his face. They travelled along his
neck and arms. Black soot skirted the uneven patches and
peppered his blond hair. There was no blood, just glistening plasma where the skin had burned away.

Mike's crew had been called to a report of an 'exploding
pavement'. It sounded like a hoax. They turned up – not
expecting much – and saw a pavement cover that looked
pretty innocuous. They removed it to reveal an equally
innocuous-looking underground electricity junction box.

Mike and another firefighter, John, had lain on their bel-
lies with their heads in the pit and tried to understand how
this could possibly be an 'exploding pavement'. They were

convinced it was a false alarm, because concrete doesn't tend to explode. Mike and Steve were standing over the pit waiting for the electricity company to arrive when the junction box exploded and a fireball erupted from it. Mike leapt backwards and narrowly avoided the flame, but Steve wasn't so lucky. There was an electrical fault and the team had misinterpreted the situation as a hoax. There were no obvious signs of danger and so they had not responded with enough caution. Had the explosion happened just a few minutes earlier, Mike and John would still have been down on the floor. It's likely they would have been killed instantly.

'Here you go, Steve. Tilt your head back,' I said. 'This will feel uncomfortable at first but it will cool you very quickly.' I lay the mask on his face. 'Just hold it there so it doesn't slip,' I instructed.

I was immediately overcome by an incredible sense of guilt. I had felt relief when another man – someone I consider a friend as well as a colleague – had been badly burned. By not wanting it to be Mike, I had wished this on someone else. I had wished extreme pain on one of my co-workers and Steve was suffering as a result.

Steve's family were the ones to receive 'that' phone call; it was their anguish, their rush to the hospital. The months of recovery, the tears, the angry outbursts, the lives rebuilt. Mike and I had avoided it, but Steve and the people who love him hadn't.

That guilt followed me around for a long time. I tried to rationalize it. I knew it made no sense, but I felt so selfish. The funny thing is that any one of us would take a burn to protect our colleagues, or someone we're trusted to rescue. This time I wasn't the rescuer, though. I hadn't even been there – not until after it happened. I was torn between the role of a responder and the role of a loved one.

Feelings of guilt were intertwined with 'what ifs'. What if the fireball had blown up a few minutes earlier? What if Mike had still had his head in the pit? What if the worst had happened? I kept replaying the alternative outcome of losing Mike in my mind. I pictured arriving at a very different scene. Each time I thought about it I felt terror, followed by another wave of relief, followed by intense guilt.

I didn't speak to Mike about it at the time. It was years before I admitted how I felt. In retrospect, I wish I had shared it sooner, but I thought it was my burden. I would lie in bed at night thinking about it, feeling like a terrible person. I would be out for a drink with friends, seemingly happy but every so often the feeling would creep in when I least expected it, interrupting my thoughts.

I know now that my initial sense of guilt was a normal reaction, but the way it haunted me less so. I shouldn't have taken so long to talk about it. However, I was afraid of telling people in case they thought that I was weak, or that I couldn't cope. I was afraid that the cynics would suggest that my weakness was a predisposition because I was a woman. I was afraid of the stigma it might bring.

I was wrong, of course, but I needed to do something with that guilt. I needed to stop it circling through my mind constantly. I needed to take control. So I decided to use the experience in the only way I could. As a driver for something good. I wanted to prevent firefighter injuries. I wanted to create a world in which neither Mike nor Steve (nor anyone else) would get hurt in our working environment. The reality is that this is an aspirational goal. It's incredibly difficult to completely eradicate injuries when we're working in environments that are, by their very nature, extreme. However, every injury spared would be a small win.

When I began to look into the matter, I discovered that 80 per cent of industrial accidents occur as a result of human error; a trend also reflected in firefighter injuries. That figure is incredibly distressing. Every single one of them is a person – like Mike or like Steve – who matters to someone, who is loved. It's not a failure of equipment, or an inadequate procedure, or a flawed policy – it's *human error*. Those injuries – so many of them – are caused by a misjudged decision or a failure to process information properly: the wrong choice in the wrong place at the wrong time. If I was going to reduce firefighter injuries, I needed to reduce human error.

I also discovered that the reason I had found it so difficult to concentrate during the four-minutes-and-thirty-seven-seconds journey to the incident was stress. Stress reduced the processing capacity of my brain and thereby reduced my ability to make decisions. It wasn't an existential experience. It was simple biology. The truth is that my chances of making a mistake were increased while I attended that incident. I was more likely to misunderstand something, or miss something, or respond in a way that was sub-optimal.

My interest in risk-critical decision-making in the emergency services had been sparked. I wanted to look at the circumstances that impact on how we operate when we make decisions that ultimately affect whether people live or die.

Since that incident, I've established a unique research group with Cardiff University that has studied incident commanders all over the UK when they've been making life-saving decisions in the field. We've used the research to develop techniques to help commanders make decisions more effectively. Techniques which are now used as standard all over the country and have been shared across international borders. Techniques that might have benefitted Steve, and that now help my colleagues and friends. It all grew from that one experience. From my guilt.

My research is about how we – the fire and rescue service – prepare firefighters to make decisions in the inevitably hostile environments we encounter in our working lives. It's about understanding how humans respond and behave in order to reduce the opportunity for human error.

Over the past decade, I have undertaken research that has painstakingly examined how incident commanders like me make decisions in the operational environment. I want to know what cues trigger the choices we make, and what circumstances affect them. Most of all, I want to know that we are doing everything we can to improve firefighter safety, to stop anyone else going through the same trauma that Steve went through, and to stop anyone else sitting at the back of a fire engine and fearing the worst for their partner or friend.

My experience – my *empathy* – drives me to push on, to continue researching and using the findings to make the working environment of a firefighter better for those whom I am proud to call colleagues, and sometimes lucky enough to call friends.

2

Wicked Problems

'He suffered from paralysis by analysis.'

Harold Geneen

THE SIRENS are piercing as I battle through the congested city streets. The traffic is virtually impenetrable and reaching the scene is near impossible. The tunnel – one of the worst places for a major incident to occur – is surrounded by towering walls of concrete. The emergency vehicles – including my own – are struggling to bypass the dozens of abandoned cars, their occupants now evacuated from the hazard zone. Exasperated, I park at the outer cordon and walk swiftly towards the scene of operations.

Thick black smoke is rushing from the tunnel and the burning vehicles stalled in the entrance are giving off clouds of acrid fumes. The pungent stench of burnt plastic fills the air, accompanied by a sweet tang that I don't recognize quite as easily, although it's inherently reminiscent of Bonfire Night. The air nearest the tunnel is laden with a heavy concrete dust. Casualties masked in soot and blood are being led away and directed towards the medical area.

Compared to the sterile serenity outside the cordon, this is a battlefield. A wounded woman, propped up by a paramedic, staggers past me. She keeps looking back in the direction of the tunnel and I expect that while she entered it with others she has struggled out alone. I've seen incarnations of this woman many times before. It catches me in my gut.

I'm approached by the current fire incident commander. I've worked closely with Lloyd for several years and he is one of my most trusted officers. He's very tall and thin; he towers over me. He is always pragmatic and an utter gentleman. I'm glad our early response has been handled by someone so focused and competent. That said, he is clearly anxious and his expression reveals signs of significant stress. But he shows no panic. I trust him. He offers a quick, confident summary of the situation. The fire has been extinguished but the tunnel remains heavily smoke-logged. There is extensive structural damage but no indication of imminent collapse. We have upwards of thirty casualties still inside and search and rescue is progressing well. We have already evacuated 115 people. The police have confirmed that the incident was caused by an improvised explosive device. A bomb. Which explains the unusual smell: gunpowder.

I look around at the dozen abandoned vehicles blocking the entrance to the tunnel; all are covered in a thick blanket of dust and dirt. Seven children have been unloaded from a minibus inside the mouth of the tunnel and are being ushered to safety. Their features are obscured by a layer of grime, broken only by the whites of their eyes. They are crying, bewildered; they don't yet fully understand the magnitude of the situation.

I notice pieces of grey concrete crumbling from the tunnel walls, presumably damaged by the blast. I want to know more about the structural integrity of the tunnel, preferably a full and thorough survey from a dangerous structures engineer. And I want to know more about the people still trapped inside: how many there are, where they are, and what is the extent of their injuries. Lloyd doesn't have the answers, though. These things take time, which is the one thing we don't have. I feel a familiar knot of anxiety in my stomach.

Lloyd fills me in on his plan and I assume command as the senior on-site officer. I radio the police commander, hoping for an update, and see her striding towards me just a few moments later. She is small, like me, with platinum-blonde hair, poker-straight and cut in a blunt bob that makes her angular features even sharper. She looks concerned.

'The situation has changed,' she barks. 'There's a credible intelligence report of a secondary, larger device inside the tunnel. Your crew are in the hot zone.'*

My heart rate increases. 'Where in the tunnel is the device?' I ask.

'We don't know,' she replies. 'But I'd guess it's intended to prevent emergency services assisting the injured, or . . .' She pauses. 'To kill as many emergency responders as possible.'

'How long have we got?'

'Our intelligence suggests detonation in the next fifteen to twenty minutes.'

'Can you stop it?' I ask.

'I wouldn't count on it. We've got to find it first.'

'How credible is your source?'

She looks at me blankly.

'Look,' I continue. 'I have twenty firefighters in that tunnel and at least thirty civilian casualties, maybe more. If we withdraw, people will die. I need to know how reliable your source is before I make that call.'

I'm craving a level of certainty that I know she can't provide.

'The intel is highly credible.' She shrugs. 'I can't share

---

* An area of danger that cannot be declared safe from the threat of a further attack.

more on the source than that – as you well know – but if I was a betting woman, I'd put money on it.'

'So, Sabrina,' Jonathan booms. 'What's your decision?'

I reach for my glass of red wine.

Jonathan and I are sitting in a quiet corner of a grand old bar. The walls are clad in a dark-stained wood that smells of tobacco smoke. The seats are upholstered in oxblood leather with worn, tufted buttons. The dim lighting feels Victorian.

Jonathan isn't a large man, but his presence far outweighs his physical being. He has a dark mop of hair, a small beard and a very kind face. He's always smiling, always pleased to welcome you into a conversation. His dark eyes twinkle. He is a trusted friend, a confidant and a wonderful mentor. And he is training me, as a coach trains an athlete, testing my decision-making to improve my overall form.

Professor Jonathan Crego has spent most of his working life exploring and trying to reduce the impact of decision inertia. This is a phenomenon in which a person's anxieties about the various things that might go wrong in a situation paralyse their ability to make a decision. It can be a significant problem, particularly in time-critical circumstances. This inertia might lead to no decision at all (decision omission) or an attempt to defer the decision, either to a later point in time or to another person (choice deferral).

Jonathan's expertise in this area is second to none and we have spent many evenings in bars discussing theoretical approaches to decision-making. We concoct ways to bridge the gap between the academic and the practical. This evening, we are the last two propping up the bar after a command conference in Scotland earlier in the day and dinner with some colleagues at the wonderfully named Weé Curry Shop. We've been coming up with exercises to elicit decision inertia and Jonathan has been talking me through an example.

'I'd like your opinion,' he'd said. 'As someone who could potentially find themselves in charge of an incident like this. Will you humour me?'

'Talk me through it,' I'd replied.

Jonathan had frowned. 'I don't want you to hear it; I want you to *do* it. Are you game?'

I'd looked at him. 'Haven't we had a bit too much wine for this?'

I knew that Jonathan had a wicked problem in mind. These are problems that are evolving constantly: the requirements are often contradictory, the scenario is always incomplete, and the aims and objectives shift. The decisions are difficult at best, and often impossible. Jonathan creates situations replete with wicked problems to challenge and test the ability of firefighters – and other emergency services personnel – to make horrible choices in horrible situations. He wanted me to be his guinea pig.

My brain ached at the thought of it. A wicked problem can't be solved, but it can be tamed. When I'm on form. Which, after a few glasses of wine, I know I'm not.

'Oh, come on. I'm not marking your performance, I just want your honest opinion. I dare you!' Jonathan chortled.

He was serious. I was a teenager being peer pressured into something I knew I'd regret. All because someone had said the magic words I have never been able to refuse: 'I dare you.'

I reached for the bottle of red and refilled my glass. I sat back, closed my eyes, and we began.

'It's 09.23 on the twenty-second of November. You've been mobilized as the incident commander to a device detonated in the Birkenhead tunnel.'

'Birkenhead!' I interrupted. 'You do know I'm in London, right? That's a hell of a run on blue lights.'

'Use your imagination. Pretend it's the Blackwall tunnel, if you must. Now stop interrupting.'

I imagined myself driving towards the more familiar tunnel in London. I listened to Jonathan's description and built a picture of my surroundings. I've driven this route hundreds of times – usually crawling along in rush-hour traffic – and I can see it clearly.

Jonathan asks questions, pushing me to describe what I see, hear and smell. It isn't just the heady mix of tiredness and wine, but I feel immersed in the scenario, like a novice all over again. The mental friction of being in a position where there is no right answer is suffocating. I am both completely relaxed and yet frighteningly panicked. I am safe, tired and warm, and yet I feel frantic, alert and clammy.

'Do you evacuate the tunnel?' he asks.

A second explosion is likely, but not certain.

I can't ignore the threat, so I must do something.

I could withdraw my crews. That will take about fifteen minutes, and will hopefully save the lives of the responders who were rushing in when everyone else was rushing out. However, if I do that I will be signing the death certificates of all those people trapped inside.

Withdrawing everyone safely – casualties included –

will take ninety minutes. At least. People are badly trapped in cars and under debris; access is difficult and specialist equipment is still on its way.

There is no good option. A wicked problem indeed.

I turn to Jonathan. 'How long will it take to complete the full search-and-rescue plan, including emergency extrication of those trapped? If I radio my operations commander, what will he say?'

'You try him on the radio but he's unavailable. No cigar. Now, based purely on what you have. Try again. You have fifteen minutes.'

I can feel myself panicking. It's irrational – I remind myself that this isn't real – but my heart is hammering.

'Talk me through it,' Jonathan says.

'I don't know how much of the tunnel we've cleared. And I don't know how many people are trapped. Or how long it will take to get them out. I have credible information suggesting that in fifteen minutes a device will explode and kill or seriously injure anyone in the tunnel. Twenty firefighters, six paramedics and at least thirty civilians. So, fifty-six lives at stake.'

'What are you going to do?'

'I could flood the tunnel with all the responders I have. We might get everyone out, but if the device explodes sooner . . .'

I think back to my experience with Steve and the fear that it might have been Mike.

'In any case,' I continue, 'briefing and directing the crew would eat up most of my fifteen minutes. They'd then enter the tunnel in complete darkness. The tunnel is one thousand, three hundred and seventy metres and they'd be carrying a good thirty kilos of equipment. I don't want to put them in the tunnel, and it wouldn't make sense to.'

'So, what's your decision, Sabrina?' Jonathan repeats.

'I could withdraw the crews. Everyone who went willingly into a dangerous tunnel, trusting me with their lives, will come out of it. And I will have fulfilled my unspoken contract with all of them. But my crews want to be rescuers – they willingly risk themselves to save others. If I pull them out, I will have failed to protect the public. And I will have forced failure upon every single emergency service worker who is here trying to save lives.'

Whatever my decision, people die. The least worst option is to minimize the number of deaths.*

'So, I ask you again,' says Jonathan. 'What's your decision? Twelve minutes to detonation.'

'I'd give the order for a full and immediate evacuation of all crews with instructions to bring out any casualty that can be *rapidly* extricated. I will be clear – and this is hard – that those casualties who require more attention should, for now, be left in situ. Crews will remain withdrawn until we receive confirmation that the device either does not exist or has been neutralized.'

'Open your eyes,' says Jonathan.

I do as instructed.

'How confident are you in your decision?' he asks.

---

* The 'least worst' option is the best option in a group of possible choices, all of which have undesirable outcomes. Because all result in a bad outcome, there is no good option, only one that is the 'least' unpalatable.

Confident. Now that's a funny word to use in these circumstances. How can anyone feel confident signing away lives?

'I'm confident that it is the least worst option,' I reply. 'Am I confident that all will be well? No. Not at all. Am I confident that my decision will result in the fewest casualties? Yes.'

These situations are fraught with 'what ifs'. What if the bomb goes off in five minutes instead of fifteen? I would wish I'd pulled the crews straight out – told them to run and leave anyone who couldn't be pulled free straight away. What if it goes off in thirty minutes? And we're all stood around waiting while civilians are stuck inside, dying? I'll wish I'd sent in additional crews. What if I find out that one of the casualties left inside was a pregnant woman, the widowed single mother of three other young children who are safe at home but now orphaned? Another three lives destroyed by the ripples. Would I have done something differently in order to save her? To spare her children from the pain of that life-changing moment?

These questions will be repeated by my colleagues, my superiors, politicians and by the families of the deceased. They too are aware that while I weigh up the options, there are millions of alternative possibilities, millions of different ripples and reverberations that may or may not come to exist. And that when I voice my decision – for better or worse – the picture is fixed.

Then, my decision will become subjected to the 'what ifs' of every newspaper headline, of every newsreader and every pundit feeding our insatiable curiosity and culture of blame. What if the emergency services were better resourced? What if they were better trained? What if there was less pressure on the NHS and more money for other public services?

Then there's the line-by-line, second-by-second forensic analysis of the inquiry post-incident. The 'what ifs' that are thrown up and scrutinized in a judicial setting. The 'what ifs' that have real consequences for improving the outcome of future events but also for affecting everyone touched by the tragedy being reviewed.

The advantage that my colleagues, my superiors, the press, the courts, the families of the casualties will have is time and hindsight. My decision needs to be made in the heat of the moment. Under the pressures of limited time and uncertainty, through the quiet whispers of self-doubt and 'what ifs' that can derail even the most hardened decision-maker. I'm making a decision as a human being. With all the flaws and feelings that brings.

Because to do nothing – to succumb to decision inertia – is by far the *worst* option. If I refuse to make a choice and do nothing, if the bomb explodes in the anticipated fifteen minutes, then I have signed the death certificates of fifty-six human beings and saved no one.

Jonathan is trying not to smile, but the subtle crinkling of his eyes gives him away. He can see my discomfort. His scenario is effective.

'Close your eyes,' he says. 'Let's go.'

I take a deep breath, sit back in my chair again and visualize the tunnel.

'What can you hear?' asks Jonathan.

The radio in my hand is crackling with voices. Those

inside, urgently evacuating. Those outside, who would far rather be inside, who are desperate to go into the tunnel and who are anxious and unsettled and will remain so until they know that their colleagues are safe.

I watch as a steady stream of rescuers and rescued stagger from the tunnel, a mixture of exhaustion and relief etched into their faces. I begin to feel more confident in my decision and start to think about the next iteration of the situation and the people left behind. What will happen when the outcome of the secondary device is known – either after it explodes or, as I desperately hope, we find out it doesn't exist or has been dealt with?

'Your operations commander approaches,' interrupts Jonathan. 'He's telling you that there's a problem, a big problem. He's clearly anxious, fiddling with his overcoat, sweating and fidgeting. He tells you that there's a fire-fighter refusing to withdraw.'

I take a deep breath. I really hoped this wouldn't happen.

'Who are they? Where are they? And why are they refusing to get the hell out?' I ask, trying to stay calm.

'His name's Matt Carnegie. He was originally deployed in a team of two. They were working on a car near to the original detonation deep in the tunnel. He's with a young girl, nine years old, trapped by her ankle in a car. It's a difficult extrication; they've been working on it for the last half-hour. His teammate has now left but Matt has refused to go. You have seven minutes.'

I know how it feels to be so emotionally invested that you cannot possibly walk away. The sense of duty and care can be so innate and become so overwhelming that you simply can't override it. I can understand entirely why

Matt can't walk away from a child, but I can't allow myself to be consumed by my own emotions.

I lift my radio to call him directly. I don't want the weight of this order on anyone's shoulders but mine.

'Matt, this is the incident commander. Are you receiving? Over.'

The radio crackles and I count the seconds. Each moment is like an eternity.

'Yes, guv. I'm receiving loud and clear.'

'Matt, you are aware of the order to withdraw immediately? And of the reasons why?' I'm reluctant to use the words 'bomb' and 'explosion' in case the radio is within earshot of the little girl.

'I'm clear,' he says. The young girl's sobbing reverberates down the radio. Everyone on the command channel can hear it. 'But I'm not coming out without her. All I need is the pedal cutters. She's got no other injuries, she's only trapped by her foot!' The radio hisses, distorting his voice. 'With respect, guv,' he continues, 'I need you to stop fucking around and send me the cutters. If she dies, I die. This is someone's little girl. She's the same age as . . .'

His voice trails off. Everyone listening fills the blank with the name of a loved one they simply cannot fathom losing.

Jonathan's voice interrupts once again. 'So, what is it?' he asks. 'What's your decision?'

I must choose between my ultimate goal – minimizing loss of life – and my instincts, which are screaming at me to save the little girl. If I choose the former, I write off the life of the girl and – hopefully – encourage Matt to leave. If

I choose the latter, I have to send in more firefighters and, if the bomb explodes in the expected time frame, they all die.

'I can't send in more firefighters,' I say. I know they'd re-enter the tunnel willingly but I refuse to put them – and their families – in that position.

'So tell me what you *will* do. What's your decision?'

Jonathan's exercises always put me in a position where I have to decide between chewing off my own hand and cutting off my own leg with a rusty knife. I want to open my eyes and say, 'No. I'm not choosing. This is ridiculous. I've had several glasses of wine and this is just silly.' But the truth is that I'm afraid. These may be fictional characters, but I want to do my best for them. And refusing to participate is simply another form of decision inertia.

Jonathan is incredibly adept at putting commanders in an almost trance-like state. He uses a steady stream of questions to force us to imagine and describe every detail of the scene. It's like a phenomenally vivid dream, except we're in control and guided by Jonathan. I suppose it's also like a nightmare: it isn't real, but the anxiety, uncertainty and discomfort certainly are.

'Decision, please,' says Jonathan.

But what if I send in the equipment and everyone dies? What if we have more time and can save the girl? What if I can persuade Matt to leave? What will that little girl feel – scared, in pain, in the dark?

I force myself to examine the facts. There's no policy for this. I have nothing to fall back on but my own judgement. All I can think of is my daughter, Gabriella. I'm envisaging her in that position and torturing myself with the image of her facing death alone. Is my judgement clear enough? Will I be able to live with my decision? Can I withstand the constant reminders in the papers, on the news, and in the courts? What if I get it wrong? My professional reputation is on the line, yes, but my well-being – my sanity – is too. Most of all – most importantly – someone is likely to die. And they are to someone what Gabriella is to me. That is the crux of this.

'Tick-tock, Sabrina. Six minutes to anticipated detonation. Get a move on.'

I open my eyes. 'I'll take the cutters,' I say. 'I'll go in myself. Then it's all on me.'

'Cop out!' Jonathan shouts, slamming his hand on to the wooden table. 'I thought you were made of stronger stuff! You go in? You leave the scene in disarray. You're the leader, aren't you? Who takes' over and steers the ship? Who follows, trying to stop you, and gets killed too? The bomb goes off. Gabriella is left without a mother, Mike without a wife. Another strike for the death toll. I thought you were trying to save lives! Not waste more!'

Jonathan is right, of course. One could argue that it is selfless to only put oneself in harm's way and the mark of a strong, fearless leader. However, it is also irresponsible and reckless. Both arguments are right, both are wrong. The same figure is definitely a nine and definitely a six, depending on which side of the table you are sitting. The truth is that my decision is neither right nor wrong – it is decision avoidance. Jonathan was right. I copped out of the tougher decision because I didn't want to make that call. So much so that I'd have loaded a single-barrel gun and spun it around, pointing it at my own head instead.

This is decision inertia. And it hurts.

'Try again,' says Jonathan.

I take a deep breath. 'I radio Matt,' I reply.

'And you say . . . ?' Jonathan says encouragingly.

'I won't commit anyone else to the tunnel.'

A black haze has claimed the area outside. I look desperately into the bleak underpass, hoping to see Matt trotting out – by some miracle – with a little girl in tow. I feel my chest constricting. There's nothing but smoke and an eerie silence as everyone looks on. Lloyd is holding his radio to his ear, waiting tensely for Matt to reply.

'The risk is too great,' I say. 'You need to reassure the girl; tell her we'll be coming back in for her. And then you need to come outside. Tell me everything you know about her location and the nature of the entrapment and we'll put a plan together to have her out as soon as we can. I need you to withdraw immediately. Over.'

'Guv, I can't leave her to die. All I need is those pedal cutters! Please!'

'Withdraw,' I repeat. 'Immediately. If you want the pedal cutters, you're going to have to come out and get them.'

I radio the police commander that I need officers on standby to physically prevent Matt re-entering the tunnel if he does emerge. I know that he'll never forgive me, that he'll never forgive himself for leaving, but that's my decision.

I hear Jonathan's voice again, softly this time. 'How confident are you, Sabrina?'

'Not at all,' I reply. 'I still want to grab the pedal cutters and go for broke, but I know it would be futile.'

'But you're decided?'

'I am. That is my least worst option and I'd be ready to live with the consequences.'

'Good.' Jonathan sits back in his chair and smiles. 'You looked uncomfortable there, Sabrina.'

'Uncomfortable? I feel like shit!' I take a gulp of wine. 'That's a horrible exercise. It's brilliant, but it's horrible. You, sir, are a barbarian!'

His smile stretches across his face.

'And that is not a compliment,' I add.

'Want to know what happens?' Jonathan asks.

I nod. I've been so immersed in the situation that I'm now invested in the outcome. I need closure.

Jonathan nods knowingly, and encourages me to close my eyes and return to that uncomfortable soup of responsibility and ominous foreboding.

'The police commander is moving her officers into position. Your eyes flicker between them and your radio. You're waiting for Matt to answer his radio. You shuffle nervously. The seconds tick on. The radio crackles and you jump, but it's a message from the command unit confirming that all the other crews are out of the tunnel and accounted for. They have successfully evacuated a further twenty-six casualties, four people are still trapped, and there are possibly others in areas you haven't yet searched.'

'Four too many, but I'm amazed it's not more.'

'You radio through to Matt. Nothing. It crackles, but he doesn't respond. He is refusing to engage.'

'He's made his choice.'

'You stand there, at the entrance to the tunnel, for eight minutes. Everyone is silent – watching, waiting. He is resolute in his decision to stay with and protect the little girl. The device detonates at 10.57. Seven minutes after the intelligence had originally indicated. Matt dies in the explosion, as does the little girl he refused to abandon. They were too deep into the tunnel: it's unlikely there'd have been sufficient time for a successful extrication. You chose the least worst option.'

Can you imagine going to work one day and finding yourself making choices that will result in the deaths of innocent, unsuspecting people? This is the position that front-line public servants are often in – firefighters, police officers, soldiers, paramedics, doctors and nurses. These decisions don't happen every day, though we know every day that they might. However, like you, we are simply human. We are limited by the boundaries of human ability and can only do what is humanly possible. But the expectations sometimes placed upon us are that we are super-human heroes and heroines who can make the impossible possible.

Jonathan's exercise was designed to elicit decision inertia. It was designed to paralyse a commander's ability to make a decision. It was designed to provoke that very human cognitive paralysis – thought getting in the way of action – in the hope that with practice we can overcome that response. It is only natural to feel concern about the outcome of a high-stakes decision. However, if you focus too much on each variation of the possible outcomes, you risk becoming tangled in a circular mental loop. Anxious about whether you're making a mistake, afraid that you might be blamed for a negative result, you choose – consciously or not – to do nothing. Because if you don't touch it, it's not your fault.

Except that isn't how it works.

Decision inertia can affect anyone, in any job or situation. In fact, the only times I've experienced true decision inertia – complete paralysis by analysis – have been away from the fire ground.

Aged fifteen, I was homeless. In my final year of high school I was sleeping rough. It was a hideous period in my life that lasted for around two years and one that I've spent many years since, trying to forget. Despite my complicated circumstances, I was determined to continue going to school and to sit my GCSE exams. And often the school meal was my only real sustenance. At that point in my life, my education was the one thing I had any control over. It wasn't much, but it was mine.

For a period, I found shelter in a derelict building. I shared the space with a number of others; some in a similar situation, others with very different circumstances. I had no specific area to call my own and nowhere to leave my textbooks. There were no lockers in my school, so I had no choice but to keep everything with me. I didn't much fancy going into care so I tried not to draw attention to myself by asking my teachers for somewhere to keep them. I hid the books in old boxes, squirrelling them away in dark corners of the building, thinking they'd be left alone. I was wrong.

The building was occasionally occupied by a man called Dick. He had a shaved head and was covered in bold, black fascist tattoos. He held some pretty unsavoury views and so, being Jewish (and relatively small and scrawny), I made every effort to avoid him. One day he discovered my books, each emblazoned with my very Jewish surname. 'Cohen' isn't exactly an easy one to hide.

I returned to the building late one night and, as I walked in, Dick sidled up to me. He smelled of cigarettes, stale beer and piss. I froze. He leaned forward. He lifted his cigarette from his lips.

'You fucking Yid,' he murmured, and held the lit end against my arm.

I wanted to scream but I couldn't give him that satisfaction. I turned and stared at him. I could smell my own flesh burning and yet, through tightly gritted teeth, I muttered, 'I'm not a Yid, you arsehole. I'm fucking Sephardic.'

I guess only a fifteen-year-old thinks a bigot will respond well to facts and detail, or have any interest in the subtleties of my North African racial heritage. He swung his arm and his knuckle caught my jaw.

'I'll kill you, you fucking Jew,' he screamed as he came for me again.

I ducked and tried to shield my face. Blood trickled down my lips. A few of the other guys sheltering in the building that night overheard the commotion and came rushing in. Two of them grabbed Dick, trying to calm him, while another, Peter, dragged me into the next room. Dick pulled free, picked up a discarded bottle and smashed it against the wall. We all knew the jagged edge of the end in his hand was intended for me.

Peter was my friend and one of the few people I could rely on at that time in my life. He told me to leave, to find somewhere else to stay. He said that I wasn't safe, that Dick was violent and known to carry knives.

I had nothing with me but the clothes I was wearing.

Peter took me by my shoulders and shook me gently. 'Sab,' he said. 'We have to go now. Do you understand me?'

I stared at him. I didn't know what to do. I couldn't decide. I know now that it was decision inertia. I didn't want to run away – that wasn't me at all – but I certainly didn't want to be attacked again. I wanted to study, to pass my exams. They were my ticket out of this mess. I needed to get my books. But Dick had a weapon.

You see how it works. With hindsight, this shouldn't have been a particularly difficult decision. But the weight of the choice was too much. I wasn't just deciding whether to stay or go, but trying to balance all the ramifications, the implications that were inextricably linked. In my mind, I was making a decision that affected every facet of my future.

Fortunately, Pete made the decision for me. We left. I'm alive.

There is no clear policy dictating an approved course of action for a scenario like Jonathan's, and we know that a set procedure isn't always helpful in such circumstances. No two emergencies are the same. It would be impossible to devise something that could be applied to every incident.

However, there are certainly times when a policy offers a framework through which a scenario can be analysed. But what if sticking to it would make things worse?

Imagine that a firefighting crew attends a reported

drowning at a lake. Onlookers report that a man had been feeding the ducks but suffered an apparent seizure and fell into the water. Crews arrive but the water is still. There are no signs of a person, let alone a live one. Policy clearly states that firefighters cannot enter water more than half a boot deep if there are no signs of life. They must call on specialist resources. Firefighters have died previously while trying to save people from drowning.* The policy is there because the risks are real. The boating lake is shallow, but not that shallow. What should they do?

It's difficult. There are no signs of life, no obvious person to save by sending someone in quickly, in a measured way, to snatch them out. Onlookers *saw* someone fall in, though. The fire commander could be criticized easily for defaulting to the policy line, for not applying discretion. However, is such criticism fair when someone is simply applying the rules laid out for a set situation?

Research has found that one of the factors contributing to decision inertia is how the decision-maker perceives their own accountability. How will their decision affect them? In the instance above, perhaps the rigidity of the policies could contribute to decision inertia. A commander may fear disciplinary measures if they deviate from the procedures, or challenges to their reputation both legally and professionally.

In a case where a person is submerged with no signs of life it is unlikely to impact on their chances of survival, though we all hope that everything possible would be done

---

* On 5 September 1999, Sub Officer Paul Metcalf drowned as he tried in vain to rescue a teenager. He was just forty years old.

to save our loved ones, even if the chances are slim. Life is irreplaceable.

As Jonathan and I wrapped up the exercise, I couldn't help but think about the tunnel.

'You're quiet,' he said.

I was surprised by how immersed I had been and how much pressure I had felt. I was exhausted. These exercises are a powerful tool to practise decision-making away from the messy incident ground, which is full of unknowns and where real lives are at risk.

Jonathan leaned forward and offered me a whisky. I shook my head, but it was too late. He slid the glass towards me.

'It's medicinal,' he said. 'And you can trust me, I'm a professor.'

3

Only Human

'It's a long journey between human being and being human.'

Anon

THE SCENE is devastating. Smoke billows from Portcullis House on Parliament Square. The sun is blocked by an acrid layer of black smog. People hover outside the building, coughing, spluttering. The structure itself is imposing. Its grey stone columns lead up to chimney stacks reminiscent of a medieval castle. If *Game of Thrones* did office blocks, this would be it. Pieces of burning debris fall to the ground from a flame-engulfed window; embers scatter like rats as they hit the floor.

This building contains office space for MPs and their staff. It is very secure and extremely high profile. Nearly one hundred metres beneath it sits Westminster Underground station. Frightened people are flooding up the stairs to ground level, adding to the chaos on the streets. There are lots of lives at stake. The pressure is high. The world's media will soon be on the scene, focusing their scrutiny on every decision made and every action taken. This has all the hallmarks of a career-defining incident for an incident commander – and how you will be defined is not always within your control.

James is fulfilling that role today and will lead the response. He will set the strategy for the incident, leading a team of more junior commanders who will help him to realize that plan by directing the firefighters responsible for actioning it. Each team member is a tiny but critical cog in a gargantuan machine. James is the chief engineer. Every decision made by every fire officer will be in line with an approach that he has dictated. He will be accountable for the outcome of this operation – the output of the machine – whatever that may be. It's an enormous responsibility and the pressure is phenomenal.

Fortunately for James, no one will die today. He is about to participate in an exercise designed to test his command skills using a simulated command unit, some virtual reality and around fifteen actors. That said, it will still define his career. James is hoping to be promoted to a deputy assistant commissioner. This is his command assessment and I am his assessor.* He can only progress to the next stage if he passes and demonstrates that he can effectively manage a major incident.

I have assessed firefighters – from junior officers to chief fire officers – from all over the country. It's no secret that I specialize in risk-critical decision-making, and some candidates are understandably particularly nervous. It's not uncommon for them to hear 'psychologist', think 'psychic', and assume I can somehow read their minds. Sadly, I can't, but I have learned to read people pretty well. I know that my presence makes the candidates uncomfortable so whenever possible I assess them remotely, sitting in a separate room with their performance projected on to a big screen in front of me.

I watch James on the screen and see him pause. He is standing in front of a door. I know that moment. Those precious seconds of quiet contemplation before you walk into the mêlée of activity and emotion. Those few seconds in which the sound of your breathing is as deafening as

---

* For any firefighters reading this (and particularly if I was ever your assessor!), please know that 'James' is not a real person. I promise that he isn't you! He is compiled from many different assessment experiences, although perhaps we can all relate to his nervous anticipation at the beginning of a command assessment.

your heartbeat. You inhale and the air travels straight to your head. You reach for the soot-stained handle of the command unit and pull open the door. The buzz of activity stops you there in the doorway and, like the heat barrier in a house fire, you have to forcibly push yourself through it.

The command unit is a large red, box-like vehicle, about the size of a minibus. It's the nerve centre of our on-scene operations, the brain of the on-site team. It's where we hold all of our information: the details of the buildings, information about who is involved, what exactly is burning and what could cause unexpected harm. The walls host whiteboards covered with diagrams and scrawled notes. A large computer screen dominates the front end of the unit behind the driver's seat. It displays a chart of the command structure – the officer in charge of each sector, the crews assigned to them.*

James steps inside and into the haze of radio traffic and noise. My screen shows everything happening in the command unit. I can see every detail on every person's face. At present, the fire is being managed by a group commander who can supervise incidents with up to ten fire engines. Now that additional crews are arriving, a deputy assistant commissioner – in this instance, James – is taking over.

---

* It can also show the message log, which documents the information passed between the incident and the control room, the feedback received from firefighters inside the building and officers on the ground. Or the key decision log, in which commanders record the rationale behind critical decisions for future scrutiny and judgement.

James needs to get a good, clear briefing from the current commander.

There are six or seven people in the unit: two command support operatives assigned to communications, two or three fire officers jostling to speak to the current incident commander (who is looking irate), and a small and unassuming police officer.

Entering the confusion and towering over the others, James looks every inch the commanding officer. He's a tall man, well built, with a resonant voice. His uniform is immaculate, his posture upright, his head held high. But his face tells a different story. His micro-expressions – brief, involuntary facial tics – and body language reveal what's going on behind the confident façade.*

'I need a brief,' James booms. 'Who's in charge?'

He pauses for a second. No one responds.

He begins again. 'Brief. Now. Please.' He looks around the unit in anticipation.

'I'm in charge, guv. Give me two minutes and I'll be right with you.' Tom is playing the role of the current commander. He's a kind man and a good officer. I've known him for several years and have always been impressed with his calm and considered manner. He exchanges a knowing glance with another colleague; they are both irritated immediately by James's overbearing behaviour.

---

* Micro-expressions often last only a fraction of a second, sometimes as a result of the individual consciously suppressing, or unconsciously repressing, their reaction. They are the emotional leakage that allows me to see what a person's really thinking and feeling. For further detail see Bibliography, p. 270.

'Well, be quick about it. I need a brief as soon as you can,' James barks.

A fire in the middle of central London is challenging at the best of times, but this is a high-profile building on Parliament Square during a period of heightened security. The scenario is designed to test the commander.

James grasps the gravity of the situation immediately. He is trying to keep his stress levels hidden, but his face reddens as he looks at the command structure drawn on the board and realizes how under-resourced they are. He's barking at his colleagues because he wants to be seen as authoritative, because he wants to hide any glimmer of uncertainty within him. However, uncertainty is inevitable in this working environment. James wants his colleagues to believe he is impenetrable, that the metaphorical bullets just ricochet off him. In this world, that's impossible.

He is overcompensating and it's making others feel stressed. More importantly, Tom has been put off by James's abrasive approach, and as a result – be it consciously or subconsciously – is delaying the essential briefing.

With the buzz of voices and radios building in the unit, James finally gets his brief from Tom. There are over 150 people unaccounted for and potentially still inside Portcullis House. The only way to be certain that everyone is out is to search the building systematically from the top to the bottom. There are no shortcuts. This is the priority. All plans, tactics, actions have been focused on this primary objective. That means nothing should be going on in the unit that isn't contributing to this goal.

'Do we know what's caused the fire?' asks James.

'No, guv, but—'

James interrupts Tom's reply. 'Any intel from CT?* Any sign of terrorism?'

'No. Nothing to—'

James cuts him off again. 'Well, get the NILO here.† We need the NILO.' He turns to the command unit operatives monitoring the radios. 'Someone get me the NILO now!'

James is considering the cause of the fire and he's right to do so. If this becomes a terror-related incident then the situation becomes much more complicated. However, he is only asking questions about potential terror activity (a bomb, most likely) and isn't exploring other possible causes.

He turns to the board to check how many fire engines are on scene. His brow furrows momentarily. I see signs of both fear and anger etched on his face.

James turns back to Tom. 'We're grossly under-resourced. Tell Control and everyone on the fire ground that I'm in charge now.‡ Make pumps twenty and I need a rendezvous point for that number of appliances.'

What James doesn't understand is that no one will ever – seriously, never – believe you're in charge just because you tell them you are. I watch the way the others on the command unit respond to him. It's not good. Their

---

* CT refers to the Counter-Terrorism Command in the Metropolitan Police.

† The NILO is the national inter-agency liaison officer. He, or she, is the point of liaison between the fire and rescue service and the other agencies, has a higher level of security clearance than most fire officers, and can therefore access intelligence that may be strategically relevant.

‡ Control are notified when a new incident commander takes responsibility.

arms are crossed and many are turned away. No one makes eye contact with him. They are completely disengaged. As a result, they are not sharing information and he is unable to draw on their established situational awareness* to improve his own. A high-pressure incident such as this can succeed or fail based on the levels of trust between the members of the team. That needs to be a priority for James at this point.

I make notes as I watch his assessment unfold. In my experience, people find it difficult to accept challenges to their performance. They often misremember the actual events completely – creating an alternative version – and I've even watched some argue with camera footage. So I've learned to bring concrete examples to the debrief.

James is ready to make decisions. He instructs a number of junior commanders to oversee specific areas of the incident. This is a sensible choice and leaves him free to set the overall strategy. I hope he is beginning to get a handle on the situation.

However, I then hear a clear demonstration that James's understanding of this incident is wrong – something that confirms my fears. Quite rightly, James started by questioning whether the incident might be terror-related, but he hasn't considered other possibilities. He's now refusing to start firefighting or search-and-rescue operations until the police have confirmed that the cause of the fire is *not* terror-related. He's concerned that a secondary device

---

* Situational awareness quite simply relates to your understanding of your surroundings, including your mental projection of how the situation might develop.

could explode, killing rescuers sent into the building. He declares that the incident is in 'defensive' mode, meaning that no one can operate in the hazard area – in this case, the building and its immediate vicinity.

The fire is getting worse, burning more fiercely with no water to subdue it, and there are still 150 people unaccounted for. If he wasn't so focused on demonstrating his authority, James might have noticed the quiet police officer hovering in the command unit, with a piece of information that would alleviate his concerns.

'But one of our officers is on the phone to someone trapped inside,' pleads Tom.

'Yes, I'm aware,' James replies. 'Coordinate the information and see if we can reach them from the outside. Maybe they can get near a window.'

Tom looks both surprised and frustrated but James fails to notice his colleague's dissatisfaction.

James is not making effective decisions. The issue here isn't decision inertia, though. It's another kind of decision trap. These traps are all flaws in the thought processes of the decision-maker; assumptions or biases or misconceptions that influence the way we make a decision. They affect how we operate, both in the emergency services and in everyday life. Essentially, they are examples of human error and, on the front line, that can be fatal.

The real concern with decision traps is that they are virtually invisible to the decision-maker. You don't know that you're falling into them – not when you're fighting a fire, not when you're having an argument with your best friend, not when you're deciding what to have for dinner.

James, for example, has no idea that he is suffering from

something called 'confirmation bias'. He is only processing information that confirms his perspective. This is a common phenomenon that occurs regularly in our day-to-day lives. For example, you might assume that all women are bad drivers. You see a woman driving at 30 mph in the fast lane of the M1 and think, 'Aha! I knew women couldn't drive!' But what about the other female drivers operating quite normally on the same stretch of road? What about the men driving like idiots? You are only looking for evidence that confirms your original bias, and so that is all you will find.

At the very beginning, when James first arrived in the command unit, he questioned whether the incident might be terror-related. He was right to consider it. The cues he'd been exposed to – the central London location, the terror alert, the high-profile people within the building – had triggered a mental model of a terrorist incident.

A mental model is that file inside your head that holds the information typical of, or relevant to, a particular scenario. For example, if I were to say that I went out for dinner last night, your mental model of a restaurant would be activated. You might think about a menu with a choice of food. You might imagine a waiter or waitress coming to take my order. You might assume I had several courses and a glass of wine. You have a file in your mind that contains all sorts of information linked to dining out.

Activating a mental model is not unhelpful in itself. I work in a challenging and fast-paced environment, and my colleagues and I often have to make decisions based on incomplete information. So having a file in your head with readily accessible and relevant memories can be essential.

However, it's important to revise your mental landscape to accommodate new information.

For instance, if I go on to tell you that I went to a self-service restaurant, had one course and a can of lemonade, you wouldn't insist on asking me repeatedly what I had for dessert and whether I chose a white or a red wine. You would revise your mental model based on the additional information I'd provided.

In James's case, however, he has completely failed to update his mental model (another tricky decision trap!). There has been other information available – a potential report from the individual inside the building, details from Tom who he's consistently interrupted, and from the police officer who he's ignored – but he has overlooked it all. He has been so focused on being in charge that he has missed opportunities to learn more about the incident.

He might, for example, have prioritized liaising with the police for confirmation that it *was* a terror-related incident, rather than simply assuming. He might have allowed Tom to give him a full briefing. He might have prioritized assimilating Tom's experience into his own understanding of the situation. He might have sent someone to question the people who had been in the building when the fire started. He could have questioned the premises' caretaker and, had he done so, he would have discovered that builders had been using hot welding equipment on one of the floors. Had he known this earlier he would have – rightly – identified this as the cause of the fire, sent teams into the building to evacuate the missing people, and put the fire out.

Every decision that a commander makes is underpinned by their situational awareness, their mental picture of the incident as a whole. This is why it is so important to keep an open mind. We operate in an uncertain world: assumptions about the incident are inevitable and can fill in the blanks usefully, but it is essential to question them regularly to make sure they still fit with the facts. If you don't, confirmation bias can flourish, with potentially life-threatening consequences.

I decide to throw James a lifeline. I pick up my radio and put a call in to Exercise Control, the team facilitating the exercise, operating the screen and coordinating the other participants. I order them to send in the NILO to confirm that this is not a terror-related incident and to reveal the real cause of the fire. I'm keen to see how James reacts to the news and how this affects his plans.

The NILO enters the unit and updates the team. I watch James. He stares at the officer, his face flushing red and his forehead sweating. He knows that both his interpretation of the situation and his consequent actions were completely wrong. Had this been a real-life incident, it could have been the most high-profile one of his career. He would have been subjected to the scrutiny of the media, the public, his peers and, eventually, the courts. His mistakes have been big ones.

However, he has an opportunity to recover the situation.

'OK,' he begins. 'Everyone, gather round. I have some good news and some bad news. The good news is that there's no sign of terrorist activity, so we're clear to start searching the building. The bad news is that I got it wrong. I didn't look further than that and I should have. We should be in there already and I need your help to catch up for lost time.'

It's not always easy to admit when you've made the wrong call, especially when you're still in the heat of the battle, and I'm impressed by James's humility. His team respond well. They are relieved that he has finally admitted what they've all been thinking but didn't feel empowered to say because of his demeanour.

James moves the incident into 'offensive' mode. This means that firefighters can now be committed to the hazard zone. James is ready to fight the fire. He orders crews into the building to tackle the blaze and rescue those trapped inside.

However, while James was in 'defensive' mode the fire has spread. He receives news that around a window at the front of the building there is cracking and spalling, both signs that the structural integrity of the wall has been compromised due to the extreme heat. There are signs of potential collapse. James is busy briefing a press liaison officer so acknowledges the information but doesn't act on it. I suspect he hasn't processed it at all. Much like when you ask someone for the time, but are so focused on whatever it is that you're running late for that you completely forget to listen to their response. Things could get worse.

James calls for a Tactical Coordination Group. This is a brief meeting of all of the agencies on scene that are working together to bring the situation under control. In this case, James will chair the meeting and be joined by the police commander, ambulance commander and representatives from Westminster Underground station and Transport for London.

James moves from the command unit to a nearby exercise room and begins to update the other agencies. He labours heavily on the early possibility of terrorism. I wonder whether this is for their benefit or for mine. Or perhaps he is trying to justify his actions to himself.

Instead of asking for updates from the other agencies, James moves on to the press strategy for the incident. This worries me. It suggests that James is still focused on rationalizing his initial actions. I think I'm seeing early signs of tunnel vision. With confirmation bias, you get what you expect because you only consider information that supports your understanding of the situation. With tunnel vision, you focus only on one part of the situation, rather than looking at it as a whole. You follow a single line of enquiry.

Tom enters with a face like thunder. He apologizes to the room and talks quietly to James.

'Boss,' he whispers. 'We need you to come straight away.' His voice is trembling.

Tom is a brilliant actor and it makes such a difference. I almost believe him myself.

'There's been a collapse,' he continues. 'We've lost contact with two firefighters.'

James doesn't react. He focuses instead on the piece of paper in front of him: the press release, the possibility of terrorism. This is a perfect example of tunnel vision. James is completely unable to process any information that doesn't contribute to the one issue he's addressing.

'Boss. We need you now,' Tom repeats. 'Please.'

James looks up. He realizes that he has grossly misjudged the situation. The problem isn't a faulty piece of equipment or a bad procedure, but him. And, more specifically, his response to stress. Stress affects the way we process information, and it is important to recognize that our processing capacity is not infinite. The more stress you are experiencing, the less processing capacity is available.

Imagine your mind as a jug filled to the brim with water. As you become stressed, some mud is tipped into the jug. First, it discolours the water, affecting your thought processes and your ability to make rational decisions. Then, if you keep adding more mud, the jug will overflow as more and more of the water – the useful space – is overtaken by mud: your response to stress.

The result is human error. James was clearly stressed from the beginning and it's why he has made mistakes throughout the exercise. Truthfully, it isn't surprising. This task is designed to be difficult. Commanders all over the country who want to progress have to pass a similar challenge. It's absolutely essential that the most senior members of our team are ready and able to take on all aspects of their new role, all the responsibility that accompanies a rank at this level.

It is not enough to be a good manager in the office, or to be very talented during the course of day-to-day duties. It is imperative that every senior commander is genuinely confident in their ability to handle a complex situation and can deal effectively with the inevitable stress that accompanies a high-profile, high-pressure incident. Command skills – cognitive skills – are critical.

On 7 July 2005, London experienced a coordinated series of bombings affecting its transport infrastructure, in particular the London Underground system. Fifty-two members of the public were killed as a result of four bombs.

Emergency responders did a lot of things right that day, but there are always lessons to be learned.

In her report, the coroner noted that the initial fire crews at Kings Cross were confronted with a dilemma that led to a delay of twenty-nine minutes. The first crew to arrive at 09.13 was a four-person crew. Procedure required an additional crew to be present before breathing apparatus crews could safely deploy into the underground tunnels. That second crew arrived at 09.42, nearly half an hour later.

The coroner recognized that neither the fire and rescue service nor any employees should be criticized for following protocols designed to keep responders safe. However, she raised a bigger question when she highlighted 'the need to balance the understandable and human urge to get involved in a rescue mission against a proper assessment of the risks involved'. She asked whether, in striking that balance, commanders – more generally – had enough flexibility to choose what actions to take.

Could they weigh up the risks in front of them and choose an action that isn't prescribed in a protocol but might be the right thing to do? Or should they always follow standard procedure?

The issue here is the subjective assessment of a risk. Can commanders judge it themselves and make decisions accordingly, even if those decisions contravene procedure? In my opinion, the answer has always been yes. Command assessments like James's are essential to ensuring that our senior commanders are not only willing and able to take responsibility but are also capable of making difficult decisions in challenging situations.

The fire and rescue service's national guidelines now recognize that commanders need some flexibility. I was part of the team that wrote them. We were keen to put the commander at the heart of the fire and rescue service's response and introduced the concept of 'operational discretion'. This allows commanders to identify exceptional circumstances where standard procedures might not apply, or occasions when sticking rigidly to the rules might be counter-productive. The commander must assess the risks and judge whether they outweigh the benefits. The procedures should provide handrails, not handcuffs.

Everything hinges on human processing. It is not just about policies or procedures (which are important for operational effectiveness), but how an individual will react and respond to a situation.

On 2 February 2005, three people died in a fire at Harrow Court in Stevenage, Hertfordshire. They were firefighters Jeffrey Wornham and Michael Miller, and member of the public Ms Natalie Close. On 3 December 2006, a fire and explosion involving fireworks at Marlie Farm, Ringmer, East Sussex, killed Watch Manager Geoff Wicker and Mr Brian Wembridge, and injured twenty other people. On 6 April 2010, two firefighters, Alan Bannon and James Shears, died in a high-rise block called Shirley Towers in Southampton. These incidents were complex and many factors contributed to the unimaginable outcomes, but coroners reviewing incidents such as these often note issues, amongst other aspects, with how risk is assessed or interpreted.

Each of the courageous people named above went to work to help others and was fatally injured in the process. I have argued for a long time that in order to truly improve the safety of our firefighters and members of the public, we need to better understand how people think and behave. This is the only way to reduce human error. If we are going to ask people to operate in conditions that are fast-paced, high-pressured and emotionally charged, then we need to make sure those people are well prepared. They need to be able to make decisions effectively, even when the pressure is on.

After Steve's injury, I started to look at the research that had already been done in this area. I wanted answers to the questions that had been spinning around in my head since the incident. It became painfully obvious that there had been little work done on decision-making in the fire and rescue service.* We were woefully behind other high-risk industries (such as aviation and medicine) for which there were hundreds of papers exploring all facets of how human factors play out in the heat of the moment. Researchers had been working closely with those industries to reduce human error for decades. I knew this was important and I wasn't going to sit on the sidelines, waiting for someone else to fix things. So I decided to plough my own furrow.

This wasn't straightforward. I had – thankfully – managed to sit my GCSEs but any further education was impossible while I was still living on the streets. It simply

---

* Professor Rhona Flin had researched non-technical skills in other high-risk industries – the cognitive and interpersonal skills needed in order to apply technical skills – such as communication, leadership, teamwork, decision-making and situational awareness. Dr Gary Klein had done work with the military and with fire commanders back in the 1980s, which informed his theory of naturalistic decision-making.

wasn't a luxury I could afford. I joined the fire and rescue service at eighteen and, when I did decide I wanted to make a contribution, I didn't have the right qualifications to make a meaningful difference. I had to understand how the brain worked before I could even think about applying that science to the fire and rescue service.

During the course of my work I received qualifications that were considered equivalent to A levels, so I could go straight to an undergraduate degree. I started by studying psychology through the Open University. As a serving firefighter, combining full-time work and studying was difficult – while my friends were out having a good time, I had my nose in a book – but I really enjoyed it. I was particularly interested in the modules that explored neuroscience. My father died of a brain tumour when I was very young and I witnessed the cognitive degeneration that came with that terrible disease. It affected so many aspects of the father I knew and loved. His personality changed. He made riskier decisions, and he couldn't manage his emotions as well. He became upset easily. Physically, his vision and balance were hampered.

I remember vividly the day he died. The twenty-eighth of July 1992. I was nine years old. Mum had helped him outside and he was lying on a sun lounger. My brother and I had just broken up from school and we were showing him our school projects from the end of term. Mine was – thrillingly – on the epoch of the Victorians. I was showing Dad a scribbly picture I'd drawn that had been up on the classroom wall, mounted on to purple card. It was too much for him. He needed to get back into the shade of the house but, being a proud man, he wanted to walk by himself. I remember him stumbling and crawling, refusing help, determined to the very last – in both the big things and the small things. He died later that night.

I find it incredible that every aspect of our conscious being – our personality, the way we perceive things or judge a situation, the decisions we make – is contained within billions of tiny cells. Activated by electrical impulses and the release of tiny neurotransmitter molecules inside that grey and white mass that is our brain, every aspect of our selves, every thought, is governed by this physical matter. That pernicious cancer didn't just take away my father's body. It took away part of his soul. And I have cursed it every day since. So the opportunity to contribute to the field of science in some small part, in some small way, was incredibly appealing.

In order to improve firefighter safety, I needed to figure out what was going on in my brain at a neural level when I was making risk-critical decisions as a commander. I was still studying in the final year of my degree when I decided I wanted to make a real and valuable contribution to reducing human error. I drew up a proposal for PhD research that would examine the mechanisms underpinning risk-critical decision-making. I knew it would be a risk – long hours in work followed by long hours of study. I could easily burn out. However, I wanted to question how commanders' experiences at fire incidents might bias the way that we respond to incidents we encounter further down the line. I wanted to know whether mistakes – such as the ones we saw from James – were a result of previous experiences subconsciously biasing our choices.

There was another hurdle. Finding a university that

would allow me to undertake a PhD part-time was tricky. Part-time PhD students are less likely to finish their research and, when students start but don't finish, it reflects very badly on the university, their rankings and ultimately their funding. Fortunately, I found the most wonderful professor, Rob Honey, at Cardiff University, who saw beyond my circumstances. Rob is an incredibly kind and capable man, and is one of the giants whose shoulders I've been lucky enough to stand on to peer over the hill.

I met with him several times before he agreed to take me on. We discussed my research and he helped to shape my proposal. I remember the day he agreed to work with me very clearly. It was a warm spring afternoon in 2008 and the sun was shining. That in itself is a rare event in Cardiff and I knew it was a good omen.

To show the confidence he had in me, Rob confirmed the university would fund my fees as well as my project. It was an amazing feeling – I had planned to take out a sizeable loan, so in real terms this made a huge difference. After giving me the happy news, Rob offered me a tour of the labs, and as we headed out, another researcher crossed our path. Rob presented me to this tall, intense woman as a 'part-time PhD student'. She looked both surprised and disgusted at the prospect.

'Part-time? With a full-time job? You'll never finish!'

I introduced myself further and shook her hand. Firmly. Through gritted teeth. As she left, Rob apologized on her behalf.

That interaction had a profound effect on me. There was no way *that* woman was going to be right. I became filled with a dogged determination, and her dismissal would be the fuel I needed when things weren't going right. Every time I was tired after a day's work but had to go to the lab, every time I messed something up (which was often), every time I felt like giving up I would think of her words. I'd had a similar exchange years earlier when I first told people I wanted to be a firefighter. They'd laughed at me. They'd said I was too small and that I'd never be strong enough. It made me determined to prove them wrong. I did then, and I would again.

Just over a year later, on the day I was due to begin my PhD research, I gave birth to my daughter, Gabriella. Professor Rob encouraged me to delay the practical research for a few months and instead focus on the literature, reading relevant psychology and neuroscience papers. During those first months of Gabriella's life, I would read papers to her as bedtime stories, just to squeeze in more work. I assumed she was simply being soothed by my voice, although as a toddler she regularly referred to a hippopotamus as a hippocampus, so perhaps something sank in.

Gabriella was six months old when I started to run the research experiments I'd spent months designing and preparing. My plan was to investigate what happened in the brain when certain patterns of behaviour are triggered by a previous experience. I wanted to know how firefighters' responses to specific cues led to behaviours and actions that restricted their ability to make effective decisions. What drives our decision-making processes? How do our experiences affect our decisions, in a good or a bad way? Learned behaviours can have many origins, but I was concerned with two primitive and fundamental forms of learning.

The first was Pavlovian learning. This is when a specific *cue*, such as a light or a sound, is paired with a specific, valued out-come and thus generates a behaviour. Ivan Pavlov was a Russian physiologist who studied the digestive system using dogs. He knew that dogs would salivate when anticipating food. This is a normal reflex (just as you instinctively snatch your hand away from something hot). However, he then dis-covered that dogs would start to salivate to stimuli that were otherwise unrelated to the food. He played sounds – such as the ringing of a bell – to the dogs before feeding them and quite quickly they would salivate whenever they heard that particular noise, even if no food was presented. They were unconsciously associating a neutral event (the noise of the bell) with the outcome of food.

The second was instrumental learning. This involves dis-covering that a specific *action* results in a specific, and valued, outcome. More precisely, it is doing something in order to achieve a desired result. For example, if the dogs had discovered that pressing a lever resulted in a bowl of food appearing in front of them, that association would be the result of instrumental conditioning. The main differ-ence between the two forms of behaviour is that Pavlovian learning is often construed as involving unconscious, auto-matic behaviours, while instrumental learning involves deliberate, planned actions. I was particularly interested in what happens when these processes interact.

For instance, imagine you used to be a smoker – one of your favourite places to light up was the beer garden at your local pub – but you've given up and your desire to smoke has evaporated too. Until, that is, you sit in the beer garden. All of a sudden – despite no longer being physically or psychologi-cally addicted – you crave a cigarette. This is an example of a cue-induced craving. The two things – smoking and the beer garden – are linked in your brain. You associate those surroundings with the joy of the nicotine rush (Pavlovian learning). And you associate the nicotine hit with the deliberate action of smoking (instrumental learning). So, when you go to the beer garden, these processes interact; a process known as Pavlovian-instrumental Transfer – or PIT for short. The surroundings act as a cue, signalling a nicotine hit, and make you crave the action that has in the past delivered that experience. This is why recovering addicts are often encouraged to stay away from places, people and actions that they previously associated with their habit.

I wanted to explore how this interaction might affect firefighters. I knew that, personally speaking, I'd developed Pavlovian associations between things I'd seen or heard at fires and specific outcomes. For example, the sound of fire engines screeching on to site increased – and still increases – my heart rate. I know that more fire engines indicate that the fire is escalating and I experience an unconscious physiological response as a result. I also knew that I'd developed instrumental associations between the actions I took – the protocols initiated or the commands given – and their outcomes. For instance, I knew that pulsing water into hot gases in a burning compartment cools those gases down and prevents a flashover.* The more times I experienced the

---

* A flashover is when an entire room or area is suddenly engulfed in flames from floor to ceiling. This occurs because the total thermal radiation from the fire plume, gases and the boundaries of the room heat all the exposed surfaces to such a level that everything catches alight at once. The room flashes over with fire.

reinforcing effect of the temperature dropping, the more likely I was to recognize the signs and continue this action at subsequent incidents. I wanted to understand how the things firefighters like me have learned affect how we respond in the future. Do our actions really represent free choice or does something about the situation – a cue – bias our responses in a particular way?

While working on these experiments, I returned to work as a station commander. I needed to be in work all day and I had on-call commitments at home. Whenever there was an incident that involved life-risk – such as a car accident where people were trapped, or a house fire where someone was missing – I would be mobilized in a response car to take charge of it. On more than one occasion I had to leave the lab to go and deal with an incident. My job was intense, being a new mum was intense, and doing doctoral research was twice as intense as everything else combined.

For months at a time across several years, I would go into the lab at 5 a.m. and run experiments until 8 a.m. when I would then go to work. At 5 p.m., I would go home to spend time with Gabriella. The sky could have fallen down in those few hours and I wouldn't have noticed because all of my attention was focused exclusively on one little being. You might call it tunnel vision. As soon as she was soundly asleep, I would hop in the car and make the fourteen-mile journey to the lab where I would continue to run experiments, often until the early hours of the morning.

When I would finally come home, Mike would be waiting up, standing over a boiling kettle and making me a cup of Earl Grey. We'd sit and talk or just hug in exhausted silence for twenty minutes so that I could wind down – just enough to enable me to sleep. Then the alarm would go at 5 a.m. and so it would continue. I had some respite at weekends – I would still have to go to the lab when I had experiments running, but at least I didn't have to go to work on top. Gabriella would come in with me and loved to clamber up to the microscopes and peer down them at slides of carefully preserved brain tissue.

It was amazing, but it was exhausting. I was planning to continue this routine for the seven years allocated for a part-time PhD, but I became determined to complete it sooner and eventually finished my research just three years later. Rob had to re-register me as a full-time student just so that I could submit in this time frame. I'll admit that part of the reason I worked so hard and so fast was to prove that insensitive woman* wrong, but mostly I was excited by my work. I knew that what I was doing would eventually contribute to making firefighting safer and to saving lives.

Over those three years I learned a huge amount about what happens in the brain when firefighters respond to incidents. Their brain processes interact in powerful ways that can bias their response. For instance, a firefighter sees a certain pattern of smoke. They have seen it before and on that occasion there was a bad outcome. At that original incident, they took a very specific action (for example, committing crews into the front door of the building with hose-reel jets). They may then associate that bad outcome with that particular action and the associations may interact. In this instance, the particular pattern of smoke might provoke a bad gut feeling, and a

---

* I actually learned a lot from her. She was an amazing scientist and the go-to for lab etiquette. She was just very socially awkward.

powerful urge not to make that response. That might not seem significant, but commanders rely on gut instinct regularly when they need to move quickly. It may, or may not, be correct when you place it alongside the other factors at that incident – the rest of the pieces of the jigsaw.

In James's case, these interactions might explain why he latched on to the idea of terrorism and then demonstrated confirmation bias and tunnel vision in his analysis of the rest of the situation. He may previously have experienced a terror incident and the cues – the central London location, the terror alert, the high-profile people within the building – may have triggered a gut feeling that made him reluctant to put firefighters inside. He needed to challenge his understanding of the situation, to question his mental model, to interrogate his learned behaviours.

Having discovered that the interactions of these brain processes *did* affect the response of firefighters, I was keen to find ways to use this information to improve training for commanders. I already knew – from working with Jonathan Crego and his colleagues – that the most effective way to combat decision traps was to focus on an operational goal, to let go of the 'what ifs' and to concentrate only on actions and outcomes.

In the situation in Chapter 2, with the tunnel and improvised explosive devices, I needed to focus only on saving the greatest number of lives possible. I had to ignore the lives that I couldn't save, and stay connected to my overall aim. Whereas in the Portcullis House fire, James needed to be open-minded; he had to focus on rescuing the people in the building and not be distracted by his own assumptions. I needed a way to make this achievable (without insisting that Jonathan attend every incident and bark directives at everyone who slipped into a decision trap!).

Therefore, after completing my PhD, I went on to the position of Honorary Psychology Research Fellow at Cardiff University, working with Rob and leading research in collaboration with the National Fire Chiefs Council.* Again, this was in addition to my full-time operational role. Over the past decade, we've made huge progress. We have driven changes to national policy and training specifications so that all commanders will learn about psychological command skills such as decision-making and situational awareness. They learn about confirmation bias and tunnel vision and all manner of other decision traps that people operating under huge amounts of pressure can fall into. They learn about the multitude of ways they might make an error in order to prevent that from happening.

As for James? It wasn't his day. The pressure was too great for him and he didn't cut the mustard on that occasion. I think he realized that for himself part way through his assessment. However, it was by no means the end of the road. Command is a skill and like any other skill it improves with practise. James is a good man who had a bad day. He got there eventually, passing his assessment the following year. He put in the hours in challenging, difficult and uncomfortable exercises. And he trained hard, so when he's doing it for real . . . it should be easy.

* The National Fire Chiefs Council (NFCC) is the professional voice of the UK fire and rescue service, driving improvement and development throughout the fire and rescue service in the UK.

4

The Jigsaw

'It's always the small pieces that make the big picture.'

Anon

ANOTHER OF my Pavlovian responses is to the vibrations of my pager. Whenever I hear it, my pulse quickens instantly. It's a familiar spike of adrenaline. When I used to do shifts at fire stations I would wake in the middle of the night to the piercing sound of bells, and my response to this buzzing is exactly the same. I glance at my watch. It's 10 p.m. and a freezing-cold November night.

My working environment can be unpredictable. One minute I'm relaxing at home in the evening, and moments later I'm anticipating being thrust into the middle of a disaster. My job these days is to lead incidents requiring eleven or more fire engines, but still, each scene could involve quite literally anything. I could be waking to a nightmare such as the Paris Bataclan, or a fire raging in a packed hotel. Or it could be a blaze in a building where only bricks and mortar can be lost. I hope for the latter. Material things can be replaced.

I need to calm my response to the adrenaline coursing through my body and be ready to make decisions. I used to jump straight in, but for the first minute or so I knew I wasn't being as effective as I could be. I would make mistakes too easily while my body was in fight-or-flight mode. So I go into the bathroom instead and splash my face with ice-cold water. My pulse slows. I pat my cheeks dry and take a deep breath.

I return to the living room – passing my daughter's bedroom door and pausing, as always, to watch her chest rise and fall as she sleeps – and sit at the table. In front of me is a small, coffee-stained notepad and a pen, my phone and my radio. Now I can begin.

I check the messages on my pager. It confirms a fire in

a large industrial unit. I know only the name of the company – Cobel & Co. I have no other information. I call Control for more details. I'm picturing a large warehouse made of brick or corrugated steel. My greatest fear at this point is a 'sandwich panel' construction which is typical of industrial units. This is a building with walls made up of highly flammable insulating foam sandwiched between two metal sheets. Fires can spread undetected through the insulation and become incredibly difficult to predict. The building then loses its structural integrity and can collapse very quickly without warning. Firefighters have lost their lives inside such buildings, including Fleur Lombard, the first female firefighter to die on duty in peacetime Britain.

As I wait for my call to be answered, I open Google Earth on my tablet and try to find the address. I know very little at this point and am making assumptions to fill in the blanks.

I'm assuming it's sandwich-panelled. I'm assuming the fire will be unpredictable. I'm assuming there will be industrial hazards. These assumptions help me predict potential problems and formulate a plan quickly. However, at the same time, I need to remain open-minded. I need to be willing to alter my mental model, if necessary.

The control operator answers my call.

'What are we looking at?' I ask. 'Are there people involved? Do we know how much of the building is on fire?' Every bit of information is a piece of the jigsaw. 'What's happened so far? How many fire engines are there? Who's in charge?'

As he responds, my pager vibrates again.

'Sorry, ma'am,' says the control operator. 'They've just made it ten pumps. We're mobilizing you now.'

'Log me as mobile to the incident. I'm on my way,' I reply.

I hang up and look at a street view of the building. It's a large warehouse. The grey, corrugated metal roof is old and tired. It resembles a rusting battleship docked in a bay. I look at the dull, cladded exterior. There's a good chance it's sandwich-panelled.

There's a van on the forecourt laden with timber. Could it be a builder's yard? I pick up my keys and gather my coat, opening the internet browser on my phone and typing in 'Cobel & Co.'. Nothing. No website. This is a big gap in the jigsaw. If I knew the type of business I could predict some of the most likely risks. For example, if it's a manufacturer, there might be materials that react with the fire. If it is a garage, there might be open pits that firefighters could fall down while searching in the smoke. There might still be people working there. It's late in the evening, but many industrial units operate twenty-four hours a day. No information has come from the scene to confirm or dismiss this possibility.

I build my jigsaw, imagining there are people trapped, and I focus on the things I do know; the pieces I do have. There was very little time between being notified (when I was first paged) and reaching ten pumps (my second page). So the fire is growing rapidly. Importantly, the response is growing rapidly too.

I reach the front door as Mike comes downstairs. As a watch commander, he understands how I'm feeling: the anticipation, the apprehension, the insatiable lure of the unknown.

'Are you off?' he asks.

I nod.

'Stay safe. Try not to chip those alloys again.' He knows how to lighten the mood, to help me relax.

I raise my eyebrows and offer him my most sarcastic grin. My response car is being returned to the leasing company in five days' time and I've just spent a small fortune having every scratch and dent removed to avoid the crippling betterment charges.

'I'll do my best,' I say.

It's dark outside and the air is fiercely cold. I pull my coat tight across my chest and tuck my chin against my neck. I open the boot of the car and reach for my blue light. I secure it to the middle of my roof, the magnetic base snapping into place and pinching the top of my fingers. I get in, switch on the lights and sirens, and I'm off.

I head north. It's going to take at least thirty-five minutes, and that's assuming the traffic behaves. I'm on the main road, trying to build some speed, to cover some ground. The entire car is shrouded in blue light. If I'm honest, this is one of the most enjoyable parts of my job. I love emergency-response driving. Making quick, safe progress along busy roads is invigorating. My sirens are blaring and I'm feeling confident as I snake between the other vehicles. Although not everyone is paying attention. I'm blocked at least three times by cars staying stubbornly in the outside lane of the dual carriageway. I have to switch my sirens from long, high-pitched wails to short, sharp yelps. Finally, the cars indicate left, clear the lane and I speed up again.

I pull into a narrow street filled with parked cars. I drive as fast as I safely can, scanning the horizon for gaps, bikes, people, and anticipating what they might do. I'm always looking for confirmation that they have seen me and my blue lights.

I listen to radio traffic and try to add pieces to my jigsaw. We're busy tonight, so there's plenty of noise, but I've learned to filter out anything that isn't prefixed by the call sign of the command unit at my incident. At first, it took a lot of concentration but, like most skills, it eventually became second nature.

I turn right out of a small, one-way street. There is a set of traffic lights ahead of me. They are red and the traffic has built up. I'll need to drive on the wrong side of the road. As I manoeuvre, a car panics and slams on its brakes. I swerve quickly to avoid it and hear the scrape of my not-so-perfect-any-more alloys against the concrete kerb. Typical. Mike will have a field day with that one.

I arrive at the cordon. Blue-and-white police tape is stretched across the road. I wind down my window, catching the eye of the police officer. I smile at her and she hurries over to check my ID.

'Evening, ma'am,' she says as she takes my card. Her eyes flick back and forth between the picture and my face.

'The rendezvous point is just ahead of you,' she says. 'There should be plenty of space.'

I thank her and park up on the side of the road behind one of the ambulances. I open the boot of my car and reach inside for my fire kit and helmet. The atmosphere is unsettlingly serene. We are a few streets away from the warehouse and the police perimeter has rendered this area desolate. It reminds me of a scene from an apocalypse film. Except in this instance the evacuation hasn't been caused by zombies but by fire. Which in my opinion is far more destructive.

Overhead, smoke is rising from the industrial unit, but I can't see any flames. I pull on my heavy fire tunic ingrained with smoke and head to the command unit. The silhouette of the building is striking. On top of ten fire engines, there are several other specialist appliances that are supporting operations. A quick tally tells me I have eighty to ninety firefighters and officers on the scene, about ten police officers and several paramedics. All working within the inner cordon. All of whose safety I am ultimately accountable for. It doesn't matter that I haven't yet taken command of the incident; I am the most senior fire officer in attendance and the responsibility is mine.

Firefighting in industrial units is inherently challenging. The conditions inside are punishing. It can be boot-burningly hot. The problems are innumerable. A warehouse can contain all manner of perilous hazards: vats of highly flammable chemicals, towering shelves of stacked goods that can fall, and dangerous machinery, to name but a few. There's also a significant risk of super-heated cylinders exploding.* I was once beside a building when one of these cylinders blew up, and I felt the ground shake as a flash of heat crossed my body. There is a reason they are treated in the same way we'd treat an unexploded bomb.

A voice bristles through my radio.

'We've requested fifteen pumps, boss.'

This means two things. First, I need to take command immediately. At ten pumps, the fire could have been

---

* Cylinders of acetylene and liquefied petroleum gas are often used in industrial processes. When heated these cylinders pose a significant explosion risk.

managed by a more junior officer under my supervision, but this is escalating quickly. Second, my jigsaw is missing more pieces than I first realized.

I run through the gaps. How many people are involved and where are they? Is there anything in the building that's going to make our job harder? What's the fire going to look like in half an hour? What resources will I need to tackle it then? What do I need to do now in order to be ready for all future incarnations of this fire? What conditions are the firefighters going into? How do I make it safer for them? And how are the commanding officers coping? Do they need more support?

Essentially, how do I ensure I have the right people in the right places at the right time doing the right things?

I also need to think about the wider impact of this incident. How might the ripples affect the rest of the city? Might it affect transport infrastructure? Are there other businesses or schools nearby? What are the economic implications for the area? All of these things will influence my strategy for tackling the incident.

The command unit is ahead of me, its blue lights flashing against the dark sky. It's the main command vehicle and everyone must go there to book in when they arrive. I reach for the door handle and step into the hub.

The unit is packed – too packed. It's impossible to concentrate effectively with this much noise. The incident commander, George, is briefing another officer. I raise my hand and signal that I need a brief. He nods in acknowledgement. I'm relieved to see that he – a group commander – has taken charge and that a more junior commander is no longer shouldering all the pressure. I pass the tally with my details to Benny, the command support operative, so that he can book me in at the incident.* I often work with him and he flashes me a huge grin.

'Evening, ma'am,' he chirps. 'Good to see you.'

'Benny,' I nod.

He knows instinctively what I need. He reaches for the keyboard that controls the main screen and pulls up the message log. I scan through the communications that have come from the incident. The messages are brief and there is very little detail. However, every particular – how many jets are being used, how much of the building is on fire – is another piece to be added to the jigsaw. I can also use the times of each message to calculate the growth of the fire – from 5 per cent of the roof to 30 per cent of the entire building in just under an hour.

So far there has been no aggressive internal attack on the fire, which might explain the rapid spread. What I don't know yet is why that hasn't happened.

George is ready to brief me. I clear the unit of all non-essential personnel. He reveals that there are several buildings on this plot. The one I saw on Google Earth – and again as I arrived – isn't on fire. The fire is in a larger building that was obscured from view. It has escalated quickly – as I assumed – and the crews are struggling to get a good water supply. They increased the incident to fifteen pumps so that they could relay water from a larger water main further away.

'Why aren't we being more aggressive?' I ask.

---

* All officers have a 'tally' with their call sign, name and rank recorded on it. It is logged by the command unit staff and is used to keep a record of everyone who is in attendance at the incident.

'There aren't people inside,' George replies. 'So putting crews in is an unnecessary risk.'

I nod but I'm not sure I agree. George is right that without lives in danger, the risk that we will reasonably tolerate is lower. However, lives aren't the only things at risk in a fire. What about the dozens of employees who will turn up for their jobs in the morning? At this rate, there will be no work for them and therefore no pay. No pay might mean not enough money to pay their rent, or the mortgage, or bills. It might mean families losing their homes right before Christmas. There is value in people's livelihoods. In my opinion, that makes it worth tolerating *some* degree of risk.

There's also the impact on the environment to be considered. Every year, industrial fires result in 136,000 tonnes of carbon dioxide being released into the atmosphere. That's equivalent to the emissions from homes and flats in Portsmouth in a whole year!*

George continues to brief me and, as always, I examine his physical responses for signs of how he is coping. This is in part to ensure his well-being, but it's primarily to see how effective he's being. I'm filling in my jigsaw based on his understanding of the situation and I need to be confident that he's processing information well. The corners of his mouth curl up as he speaks to me. His brow furrows when I ask a question. He hesitates and stutters when he answers, even though the answers themselves are clear and assured. He is anxious and, understandably, stressed. He may be

* According to 'The Financial and Economic Impact of Warehouse Fires', research commissioned by the Business Sprinkler Alliance and conducted by the Centre for Economics and Business Research (2014).

processing information less effectively as a result, so I'm keen to double-check his understanding of the situation.

'Thanks, George,' I say. 'You've done a great job. I'm going to take over and I want you to be my operations commander. Let's take a tour of the incident ground so I can get my bearings.'

George nods and exhales loudly. A recce will develop my own situational awareness, but it will also encourage George to decompress and calm down before he takes on his new role. It's as much for his benefit as mine.

I turn to Benny. 'Please let everyone know that I've taken command.'

George and I head to the scene of operations. I'm not convinced we should be operating defensively, but I need to see the fire and confirm this before I change tactics. We move quickly. There's a good chance we'll get a visit from other principal officers in due course and I want everything in place before they arrive, so that whatever they build is on solid foundations.

In front of us is a structure with nearly half of its roof engulfed by flames that crackle as they storm through the building. The heat stings even on an ice-cold November night. Dozens of firefighters are rushing around, repositioning lengths of hose and moving equipment. I approach the sector commander who is responsible for overseeing the tactics in this area. I want to make sure that my understanding of the situation – gleaned from George's – matches theirs. It can be so easy for messages to get lost in translation, especially when there is so much information to process. He confirms George's account and I begin to feel more confident in how George has handled the situation.

I finish my tour of the incident ground and return to the command unit. My jigsaw isn't complete but I'm convinced that the pieces I have are an accurate reflection of what's happening.

I have a small window of opportunity, a chance to prevent this plot from turning into an empty car park where a building once stood. The fire is currently spreading through the roof – I'm concerned about the sandwich-panelled walls, though. At the moment, they are unaffected but I have a plan that might prevent further damage.

'George,' I say. 'We need to move to an offensive mode of operations.'

George's face drops.

'I understand why you went defensive,' I continue, 'but I think there's another option. I want you to cut a fire break in the roof.* The fire is here . . .' I point at the hand-drawn plans on the board. 'In the west side of the building. And if you look here, on the east side, there's a solid-brick partition wall separating the two units. If we cut the fire break to one side of the wall, the fire won't be able to spread internally beyond it or externally through the roof. It should buy us some time.'

George nods.

'Track any spread through the walls using thermal scanning.† Hopefully the fire won't spread through them.'

---

* A fire break is a gap cut in the roof of the building, which will stop the fire from spreading past that point.

† Where a thermal imaging camera is used to track the temperature of the outside walls of the building. Elevated temperatures can indicate the fire is spreading within the walls.

I check that George is clear and that he has enough resources, and then send him off.

Although I know what the situation looks like from the outside, I have very little knowledge of what's actually happening in the building. The breathing apparatus teams will soon be inside, so I'll be receiving information from them. However, it will be delivered to a sector commander first, who will interpret the details and then relay them to George, who will pass them on to me. I will be seeing the situation through the eyes of several other people. This is why, as an incident commander, it is so important to keep questioning the information, to keep challenging and checking any assumptions, so that you don't rely on a distorted message. It would be great to receive the information directly, but for an incident of this scale it would be overwhelming, and you'd be so busy receiving updates that you'd have little time to enact a response. So it's key to delegate control of specific areas.

'Can I get some updates?' I ask Benny.

'Do you want me to get George, boss?' he replies.

'No,' I say. 'A radio update will be fine for now.'

It's reassuring for me to see the whites of someone's eyes as I receive a briefing, but I know how frustrating it can be to be called away from your operations to give someone more senior comfort that we're winning.

I hear the familiar static crackle of the radio as Benny starts to gather information.

I now know that Cobel & Co. is a joinery with stores of highly flammable varnishes and paints on site. Thankfully, they are housed in a unit that is currently unaffected by the blaze. Benny confirms that the fire break is nearly complete. We're not winning yet but the fire isn't getting any worse. I finally have a good idea of what the here and now looks like, so I can turn to the future and address the questions that have been flitting through my mind since I arrived. I'm always in this intensely inquisitive mode; sometimes it's useful, sometimes it's simply exhausting. It's like a permanent state of hyper-vigilance.

I'm used to it, though. I've been this way for years. I think it started when I was much younger. After losing my father, most of my teenage years were spent as a scruffy kid with nowhere to call home. There were times when I came to harm, and quite a few times where I only narrowly avoided it. It was shit but it made me sharp. I was always looking out for anything out of place, anything that might spell trouble. It's turned out to be a real advantage.

I glance out of the window of the command unit and notice that the wind has picked up.

'Benny, what direction's the wind?' I ask.

'North-westerly now, boss.'

'What *was* it?'

'Easterly.'

'Shit. I'm going back out.'

I head back towards Sector 1 – the busy part, the burning part. The spread through the roof space has been contained, but the wind is now blowing in the opposite direction. There's a chance the fire might jump to the garages behind us, garages full of industrial cylinders and combustible materials. I can see the burning embers of the fire dancing in the sky and landing dangerously close to the roofs. If the wind changes direction again – which it may well do; British weather is notoriously unpredictable – there's even more of a chance that the embers could jump the fire break and catch the rest of the unit, including the chemical store. I need to be prepared for both eventualities.

I head back towards the command unit. On my way I notice a man standing just behind the cordon tape. His eyes are red and swollen; his hands are clasped around his cheeks. I recognize the shock, the anger, the confusion. This is a man who has lost everything – his plans, his hopes, his future.

'Benny,' I call, as I reach the unit. 'Do we know who that guy is?'

'No, boss,' he says, poking his head outside. 'I haven't seen him before.'

'Have we managed to get hold of the owners?'

'Still trying. The number we have goes straight to the answer machine.'

'Can we get someone over to that guy, please?'

'On it, guv.'

I need someone who knows these buildings, someone who knows what they were like before they were chewed up by flames. If this man is who I think he is, he might be able to explain the dull, lifeless line-drawn plans that we've been using. He might know if there's anything in there that can cause us harm, anything that might not have been immediately obvious, anything that can help me keep my firefighters safe.

At this point I decide I need a face-to-face conversation with George. I need to change the strategy to cover both eventualities, so that we have next steps should the second building catch. George has been quite anxious throughout and I need to know that he'll cope with this change in tactics. He arrives at the unit and we gather in front of the boards.

'George, tell me what you know.'

As always, I need to be sure that our jigsaws match.

'Sector Four's looking good. We used the aerial ladder platform* to cut the break in the roof, and it peeled back really well. It took us a while, just because it's a fiddly, shitty job, but the crews did good.'

'Have you got water going on it?' I ask.

'I've got a few jets. I want to pour water through the roof, but we've not got enough. The wind isn't blowing eastward at the moment, so the jets might do.'

'I'm worried about the wind,' I say. 'It's blowing north-westerly but that's towards Sector One. The garages are extremely close, and if we lose that sector we'll be boxed in.'

George's face falls. He's realizing that he hasn't considered the ways the situation might progress. He's been so focused on the here and now that he's missed the potential development. I also know that he's risk-averse and I need to empower him to make decisions, to be confident.

I soften my tone. 'Go and have a look. Get a plan in place and then report back. We need protection on those garages – firefighters with jets or a ground monitor.† Take

---

* A vehicle that has a long arm with a platform attached to it. It's designed to deploy an elevated stream of water, or to provide a method of rescuing people trapped at a height.

† A small, metal device designed to be attached to a hose and left unattended to deliver a stream of water on to a fire. They can be set up quickly and left in places where it might be risky to leave a firefighter.

a look and decide on the best plan. We should get access to the garages and take out any cylinders, just to be safe. I'll try to get more water.'

George is staring at me. He's worried that he's missed something.

'You're doing a good job, George.'

He manages half a smile.

We spent the whole night fighting that fire. We were looking constantly for the latest pieces of information, always trying to find another piece of that puzzle. It was cold, dirty and wet. Everyone on that fire ground dug in and worked and worked until they were either relieved by fresh crews or the job was done.

We saved the garages and most of the east side of the building. Fortunately, a significant part of the joinery could still operate, so many jobs were saved that night. The man with swollen red eyes – the owner, as I had suspected – fell to his knees with relief when he realized his world had been saved, at least in part.

It could have been much, much worse. Our success was testament to the firefighters who worked so incredibly hard. And to George. He performed well under overwhelming pressure and pushed himself far beyond his comfort zone. I was incredibly proud of what he achieved.

Jigsaws such as this are significant to all aspects of life. Every decision you make and every action you take in every part of your day-to-day existence is based on what you believe you are dealing with. If you start to see things in a particular way, time after time, it becomes your lens. Your way of making sense of the world.

For me, this is significant. My lens was defined well before I joined the fire and rescue service. As a young girl of fifteen or sixteen, entirely alone, I was incredibly vulnerable. And I knew it. I saw danger everywhere. The only way to ensure my safety was to notice everything. All the time.

When I was spending my nights in a derelict building, I wasn't the only one there. As my run-in with Dick demonstrated with razor-sharp clarity. Each night, I would find a corner and then check that there was a nearby escape route – a window, a door. I would set up traps in case I had to run and wanted to hinder my pursuer's progress. I'd stack paint cans – that I'd found in a skip – near a door, so I could pull them down behind me. I piled newspapers at various points on my way out, things to throw at someone and slow them down if I needed to. When I finally settled and closed my eyes, it was always with a heavy piece of wood tucked beside me; a weapon, just in case. There were a few occasions when I was glad that I did.

I spent a lot of time creating dangerous scenarios in my head and built mental jigsaws to deal with them. Every potential threat was another piece. I needed to predict horrible situations – and anticipate how things might unfold – so that I could avoid them. Every time someone approached me, I had to decide whether to trust them. I had to judge the validity of what they were saying, their intentions. Were they being genuinely kind or were they trying to reel me in to then exploit me? I learned to second-guess everything and everyone. All the time. It was where I first learned to read people, long before I had any psychological training.

Back then, in my state of hyper-vigilance, I was always looking for another piece of my jigsaw, which I'd then test and test and test again. When I thought I knew what my jigsaw looked like, I'd pull it apart and check it once more. It was crucial to do so. My safety depended on it.

While I'm not as intense these days, some of those old habits are ingrained in me. I'm never satisfied with the picture I have. There are times when this can be a powerful asset. At this particular fire, for example, it ensured I second-guessed George's decision to adopt a defensive mode of operations. It made me question the wind direction. It saved the building and people's livelihoods. However, there are other times when I have to make a conscious effort to stop myself before it becomes all-consuming. I still have a tendency to overthink things and assume the worst. It makes me unnecessarily anxious.

We all compile mental jigsaws, all the time, without realizing. They form the basis of every decision and choice we make. They are essential. Though to make the best use of them, we shouldn't rely on them 100 per cent or leave them unchallenged.

You see, a piece of the jigsaw might be something you can see, hear, smell or taste. For a firefighter, it could be a flame peeping through a roof, the colour of smoke, the sound of a fire crackling or the taste of soot. These are the most valuable pieces – the most reliable – because they are built on your own direct situational awareness and you can trust their accuracy. This is why I insist on seeing the fire myself, smelling the smoke, feeling the heat. However, there are never enough of these solid, irrefutable bits of information. In a jigsaw containing one hundred pieces, there might only be ten of this type.

So I rely on information from other people, which isn't always as dependable. James and George are great examples. If James had told me the Portcullis House fire was a terrorist incident, would I have taken that information at face value? If I had accepted George's defensive approach to the warehouse fire, and not insisted on gathering my own information, how might the situation have evolved differently? As I add pieces to my jigsaw, I keep those obtained from others – those based on my interpretation of their interpretation – slightly fuzzier. If I don't have the same confidence in that part of the puzzle, I interrogate it as the incident progresses. Does it still make sense? Does it fit with the other jigsaw pieces? Is that bit of detail incorrect or is the overall picture that I've built wrong?

There are some people whose reports I trust more than others. I know that those people gather reliable, unbiased information and check it consistently. On the other hand, I know that they are only human. I usually judge their information based on how they are responding at the time. Their micro-expressions, their body language – not just what they say but how they say it. All of the clues that indicate how well they are managing the pressures of the incident.

When I have pieced together information gathered directly and indirectly, there are still gaps. And the human mind doesn't like to leave gaps. So, more often than not, your brain will try to fill them in for you. Not with information that you have to hand, but with information it has filed away. Your mental models. This might be knowledge of this type of incident, or previous training, or even memories of a similar experience. Your brain fills in the voids and completes the picture.

This *can* be really effective. But not always.

Sometimes, your mental model will be based on direct experience or reliable information. However, in the absence of these things your brain will fill the gaps with assumptions, which will be things that make sense given what you *think* you're dealing with. Remember the pram I found in the terraced-house fire in Chapter 1? It led me to assume there was a child in the house. This wasn't information I uncovered myself: I didn't see a child or hear a baby wail. And no one had told me that there was an infant in the house. However, there was a cue, a clue, a trigger that uncovered a jigsaw piece. Of course, in that instance, the piece was wrong.

That said, if you wait for concrete information in fast-paced and unpredictable situations, you might miss the opportunity to act, and could lose lives as a result.

The difficulty for me is when I have only a few pieces built on my situational awareness, say one or two. Generating an accurate jigsaw at the scene of an incident in this way is challenging enough, but there have been many occasions when I have had to do this remotely.

For several major incidents, such as the Westminster terror attack and the Finsbury Park terror attack, I was leading our Brigade Coordination Centre, a major incident room located elsewhere.* I was therefore away from the scene without the luxury of solid, direct, on-the-ground information. I've been in many similar major incident rooms, both operationally and through my research, which has taken me all across the country. Your focus is always broad, on a whole county or city, and while there is no operational need to be at the scene, you certainly need to understand the situation so that you can anticipate developments and coordinate people, resources and information accordingly. All while maintaining a fire and rescue service for the rest of the city in case of another incident.

It can be really hard to do this without a clear picture that has been developed on the ground. Away from the scene, you can't rely on your senses. You might have a live feed of police helicopter footage but there's no detail. You might have CCTV footage but, again, no detail. There might be a live news feed on a screen in front of you, with a real-time image of the scene. However, it's not the same as experiencing the incident first hand.

At a major incident room, we can also find a huge amount of information on social media, such as pictures and videos uploaded as soon as the incident begins. Take, for example, the haunting images of the Bataclan attack. These were shared around the world and, like many others, I watched as the horrific events unfolded online and on the news.

---

* Because of the sheer scale or complexity of major incidents, we sometimes need to introduce an additional layer of support. The role of the major incident room is to ensure that the scene of operations can be resourced and that information and logistics can be coordinated accordingly.

Even though I was hundreds of miles away, those pictures showed me the scale of the attack, the sheer number of people involved, and the videos took me to the heart of it. We all felt, watched and heard the terror as people fled, frightened and bleeding, clutching their friends and loved ones, desperate to keep them in sight. We all felt the huge emotional pressure that the emergency services were experiencing, and the stress that the decision-makers would have been feeling is unimaginable.

Remote from the scene, my concern is always the reliability of the images I'm seeing. You don't always know when or where the images were taken, and they are very often uploaded minutes, if not hours or days, later.

With only sketchy details available, most of the information at a major incident room must be acquired through other people. Updates from the scene, for example. Updates from the commanders at the Strategic Coordinating Group (SCG).* Updates from other responding agencies. The common theme here is that all of the information is coming from other people. Fuzzy jigsaw pieces at best.

It is important that everyone – not just those in the major incident room, but those on scene and the commanders at the SCG – is working with the same version of the jigsaw; not only the same pieces, but the same overall

---

* The strategic commander (or the 'gold' commander) in overall charge of each emergency service is responsible for formulating the strategy for the incident. Each strategic commander has overall command of the resources of their own service, but will delegate implementation of decisions to their respective tactical-level commanders. When strategic commanders from respective agencies meet, they are known as a Strategic Coordinating Group (SCG).

picture, which is much harder than it sounds. We're all looking at the same information but through the lens of our own experiences, training and knowledge. We're all making our own assumptions. So even though we have similar bits of information, they might add up to a different picture.

Imagine giving identical boxes of Lego to two different children. It's likely that they would put them together in a different way (regardless of the instructions) and that each model would be wildly different as a result. In a similar way, it's no good at all if commanders have the same pieces of the jigsaws, but have created entirely different images!

There are ways to minimize this. I like to ensure that the major incident room is the sole conduit for the distribution of information to anyone working remotely – the keepers of the common operating picture for fire. This means that personnel on the ground aren't being pestered constantly by the different teams, and that we all share the same picture. However, it means that if we get it wrong in the major incident room, everyone relying on us for their picture of the scene has it wrong too. This is why it's important, as the incident progresses, to regularly stop the room – pausing all of the activity, all the buzz – to get an update from everybody involved and revise the jigsaw accordingly.

There are some commanders who don't subscribe to this method. Some prefer to keep their jigsaws private, distributing information to other teams only as and when they feel it necessary. It ensures that they are in complete control and no one else in the team has enough detail to make their own assumptions. They would argue that it is a protection mechanism against creating several versions of the truth. Personally, I think everyone needs to understand how their remit affects the bigger picture. It's not gratuitous or unnecessarily time-consuming. It's critical.

Smaller, less-publicized incidents can be just as challenging as the huge, televised, resource-intensive ones. For instance, if a fire affects commuter rail lines or a key tube station, firefighters might struggle to get in for their shifts, and if I don't have enough staff, I can't use that fire engine. I need to pre-empt this succession of events and make alternative arrangements so that the service can continue. Fires don't quit because it's inconvenient. Neither can we.

The warehouse fire at Cobel & Co., and each major incident room that I've run, required advanced situational awareness. The jigsaw needed to stretch to the future. It isn't always easy to find that kind of headspace given the dynamic, fast-paced and high-pressured situations in which fire commanders often find themselves. This is especially true for initial commanders – watch commanders like Bert from the house fire in Chapter 1 and my husband, Mike – who are first on scene with so many things to do and very little time.

Imagine arriving at a house fire. Every window is spewing raging flames like the open doors of a furnace and black curls of smoke billow violently from every crevice. Everyone is screaming at you to do something, screaming that there are people trapped inside. Everyone expects you to know what to do, including your crews, who are trusting you with their lives.

You need to be quick. You respond instantly, and every time something new happens you move your attention to that next crisis. I researched this kind of scenario, unpicking how commanders respond to new challenges, interpreting their actions and measuring their situational awareness. I discovered that it's not uncommon for commanders – especially those who are there at the beginning – to make risk-critical decisions based on a very simple level of situational awareness and an incomplete jigsaw. They operate in the moment, without the time or capacity to think further into the future.

As a result, they were failing to anticipate what might happen next. I discovered that they were responding to each piece of the jigsaw individually, independently, as it arrived, and without trying to slot it into a bigger picture.

This is understandable. Remember, there are flames pouring out of this building, and these commanders are doing their absolute best in phenomenally difficult circumstances. They continue to do their best, again and again, shout after shout. However, the more often they manage incidents in this manner, the more ingrained the behaviour becomes. They progress through the ranks and are so conditioned to working in the here and now that they struggle to adopt new methods for advancing their situational awareness. Even when the working environment has changed, it's easy to slip back into what you know.

I wanted to develop techniques that commanders could use to increase their situational awareness successfully. In the course of my research programme, Rob Honey and I tested incident commanders in a range of situations, encouraging them to consciously prompt themselves to consider what they were expecting when they were about to action a decision. For example, when they tasked their crew to put out a fire or rescue a child, to anticipate how the situation would unfold. It sounds relatively simple, but it's a learned skill. It essentially imposes the conscious on to the subconscious, challenging assumptions and gut instincts in an incredibly effective way. The research we published demonstrated that this method did increase levels of situational awareness – dramatically. Just by consciously prompting yourself to think ahead.

Not just fire commanders can benefit. The technique can also be applied in other industries where decisions are made in high-pressure, fast-moving situations. Take, for example, the field of surgery. Decisions often need to be made quickly and confidently, and sometimes without all of the information to hand. Research has shown that these swift decisions made by surgeons are often done so intuitively, which is just how fire commanders operate. A technique that would prompt surgeons to consciously challenge their assumptions could be incredibly beneficial for patient safety. The practical applicability of such a technique in the field of medicine, along with other factors, is something Rob Honey and I are currently exploring with colleagues from the School of Medicine at Cardiff University.

Or take everyday life. How often do we make decisions based on the here and now without thinking about the ripple effect our actions might have? How often do we assume something is a good idea without thinking about the bigger picture, the implications? I find myself using the technique when I'm about to go for another piece of fudge cake and break my diet for the thousandth time . . . because what harm could one more slice do? Well, there's 900 calories in a slice (I like a really big slice), and it would take a

10-mile run to burn it off. Suddenly, the extra slice isn't feeling worth it any more!

I also use the technique to persuade myself to get up for a run when the warmth of my bed is much more enticing. I think about going for that jog and how much better I'll feel for the rest of the day, and compare that feeling with how groggy I'll be if I waste another morning lazing around in my pyjamas.

Admittedly, there have been times when it hasn't worked out so well. I once (or twice) missed the opportunity to apply it to leaving the pub when I told myself I was going to, and to saying no to that extra glass of wine. Although I blame the several glasses of wine I'd already had that evening for my inability to use it successfully!

When you do use it, however, the possibilities are endless – do the food shop now rather than later, pack the kids' bags the night before school to avoid the morning rush and meltdown, even though you're so, so comfortable vegging out in front of the telly. We can all benefit from extending our jigsaws – our situational awareness – before we make decisions. You might just change your own mind.

5

Trust Your Gut

'The intuitive mind is a sacred gift and the rational mind is a faithful servant. We have created a society that honours the servant and has forgotten the gift.'

Bob Samples

YOU MAKE hundreds of decisions every day. These might be big things – such as whether to get married or divorced, where to live and what to name your child – or very small things – what to have for lunch or what to wear. For every decision and every resulting action there is a response, or a reaction. Every choice has a consequence.

In my line of work, these consequences can be the difference between life and death. They can be the difference between someone walking home and someone being rushed away in an ambulance. They can be the difference between receiving a phone call that says, 'You won't believe what's just happened' and the one that says, 'You need to come to the hospital now.'

So how do we make those decisions? What drives our actions? Are they based on facts and evidence? Or instinctive gut responses? This is a question over which I have obsessed. How do we – using the mass of cells and electrical impulses within our brains – make sense of our environment, interpret how we feel and ultimately respond? I have long argued that it is only by understanding how risk-critical decisions are made, in the moment and in high-stakes situations, that we can learn to do it better.

I know from experience that circumstances can affect the way we react even to the most simple choices. Before you judge someone else's decision, consider the moment in which they made it. What would you have done in that situation? What would you have done without the benefit of hindsight? Would you have responded in a calculated, analytical way? Or trusted your gut?

*

The open road is peaceful – serene even – lined by tall, sweet conifers. Their green-tipped branches sway as the sun filters through the leaves and dapples the road. It's an image that would look very much at home on a postcard.

I, however, am seeing it from the front cab of a fire appliance barrelling along a mountain road at 60 mph. The engine is screaming and the atmosphere inside the vehicle is tense. The tranquillity of the setting does nothing to settle us.

We've been mobilized to a road traffic collision involving multiple vehicles. We received a call from a person at the scene, who gave very few details. They were likely terrified. They may have been involved in the crash; they may even have been injured themselves. Giving calm, accurate information when at your most frightened, stimulated by adrenaline and against the deafening sound of your own heartbeat, is not an easy task. However, the emergency services are entirely reliant on that information to get to you.

A second truck will meet us at the incident, but it is at least fifteen minutes behind us. It's one of my early postings. I've been serving for about six years and I am the junior officer – the watch commander. I will take charge. My jigsaw is scant. All I know is that people are trapped in their vehicles – possibly with life-threatening injuries – and that they should be somewhere along this road.

So we do our best. I'm not local and I'm new to the station, but my crew know this area intimately.

I turn towards the back of the truck.

'Does anyone know how long this stretch goes on for?' I ask, worried we're on the wrong road.

'Maybe it's a Mickey, boss,' replies Alex, one of my firefighters.*

'I don't think so. I've got a bad feeling,' I reply. I know intuitively that this will be a serious incident. One that will challenge me and my crew. I have attended numerous collisions involving multiple vehicles, and I know that cars have no sympathy for their passengers when an impact occurs.

I'm about to radio Control to ask for clarification when we take a sharp right.

'Stop!' I yell. We screech to a halt. Then that familiar, desolate moment hits – a fraction of a second of anxious anticipation as I blink back the devastation in front us. We are just metres from the first wreckage.

Tangled vehicles are scattered across both lanes. Several injured people are lying in the road, their crumpled bodies strewn across the tarmac. They've either crawled from the wreckage or were jettisoned on impact. Even through the muffling of the windscreen, the dull groans, the piercing screams and the agony of loss are deafeningly clear.

I reach for the door and task my crew to address the most pressing issues.

'Alex, Tom, get the equipment. Sandy, the trauma kit. Start assessing casualties. I want to know how many, where and how bad. Grumps, get the hose-reel off and isolate the

---

* No prizes for guessing that 'a Mickey' relates to 'taking the Mickey' and refers to a hoax call. I'd always attributed this to Mickey Mouse, until someone pointed out that it's actually cockney rhyming slang for Mickey Bliss . . . I'll leave the rest to you.

cars' batteries.* Then I want an ETA for the ambulances. Radios on. Channel one. Quick as we can.'

My boots hit the floor with a thud and I slam the door shut. I pull on my helmet as I survey the scene. I need to understand the full extent of the trauma in order to formulate an effective response.

I'm drawn to two cars mangled into a single wreckage. They are so badly crushed that I can't determine where the first one ends and the second begins. Fingers of jagged metal protrude from the vehicles. A matted, bloody mop of hair lies divorced from its body on a dashboard, visible through what was once a windscreen. I'm confident that no one could have survived such a trauma and so I move on.

I cross to the other lane where a green hatchback is lying on its side. Sandy, my crew commander, is talking to someone through the windscreen, assessing how many are trapped and how badly hurt they are. Sandy has that area under control and so I walk the perimeter of the scene, building my jigsaw. We have limited resources and I need to decide which cars, which injured people, which lives are the priority.

A fourth car, a white SUV, lies nose-down in a ditch. A tall, well-dressed woman is standing on the side of the road. She is wearing a knee-length beige coat with red stains stretching down her right side. There is a lot of blood and I don't think it's hers. She's waving her arms frantically at the SUV and screaming a man's name. She is desperately upset but fear is making her hysterical. It's

---

* This removes any potential ignition sources from sparks from the battery that could start a fire. We get a hose-reel ready in case of a fire.

beginning to have an impact on the others who are trapped and injured – panic is spreading like emotional electricity through the air.

I have only four additional crew members and they are currently dealing with four vehicles and approximately ten people. It's not enough. I need to avoid being drawn in, so I can deal with the scene. I turn to a young man sitting quietly on the side of the road, dazed and in shock.

'Are you OK?' I ask.

He nods.

'Are you hurt?'

He shakes his head. 'No,' he replies. 'I wasn't in the crash. I was driving behind them. It was me that called you.' He is a slight man, I'd guess in his late thirties, with greying hair and bright blue eyes that drill into me.

'Can you help me with something?' I ask.

He looks up and nods eagerly. In a situation like this, most people just want to be useful. Sometimes it can be a hindrance, as overzealous bystanders, humming with adrenaline and a desire to help, fail to generate anything other than chaos. Sometimes, however, with the right disposition, onlookers can make a vital contribution.

'What's your name?' I ask.

'John,' he replies.

'OK, John,' I begin. 'I need some help with that lady over there. She's very upset and I need her to stay calm. Can you help?'

'I'll try,' he replies.

We walk towards the woman and I introduce myself calmly before presenting her to John. He envelops her immediately, wrapping his arms through hers. John might have walked away from this incident with some emotional bruising, some anxiety, some level of fear. However, sharing this woman's panic will make the experience that much more intense. Neither will forget it. But there's no alternative.

I take a deep breath and carry on.

We are working to the 'golden hour' principle. Ideally, we will have all the injured people in hospital within an hour of the collision occurring. Emergency treatment is most effective in this first hour and we're about fifteen minutes into it already. We're fifteen minutes away from the nearest hospital so we have half an hour to get those still alive cut out of their cars and into ambulances. We need to work fast.

I beckon Sandy while the rest of the crew finish preparing our equipment.

I know that there are four cars involved. Sandy rushes towards me and confirms that there is no one alive in the two that are knotted together. The third – the green hatchback lying on its side – had five people in it. Two were thrown through the window as the car rolled and are lying in the road, hurt. They weren't wearing seat belts. Another – a man – is unconscious in the back of the car, and the driver is trapped but conscious and breathing. The front passenger climbed out before we arrived and is tending to the two passengers jettisoned in the middle of the road.

'How bad are they?' I ask.

'Four are pretty bad, but stable. I'm worried about the unconscious guy in the back, though. He's struggling to breathe and I can't tell if his airway's clear.'

'The car in the ditch?' I continue.

'The woman climbed out, but her husband is still inside and their two kids are stuck in the back.'

'And?' I ask.

Sandy shakes his head. 'The husband's gone.'

'Are you sure?'

Sandy nods. 'One hundred per cent. He's been impaled by a fence post.'

'And the kids?'

'They're both conscious and breathing. But too quiet for my liking. It might be shock. They need to come out so I can be sure.'

My instincts are telling me to get the children out – I don't want one of their formative memories to be cowering and trapped in a car with their dead father – but to do that we have to cut off the roof. My priority has to be those who are most injured – I don't have enough crew to do both cars simultaneously. In an ideal world, I'd sit down with a notepad and pen and think about the pros and cons of each possible step in order to decide on the best course of action. However, that's not the world we work in. I have to react instinctively; I need to trust my gut. I know that I can't prioritize the children yet.

'OK,' I say loudly so that two of my crew, Alex and Tom, can hear me. 'This is the plan. We need to start with the hatchback. We need to get the man in the back out as soon as possible.'

Alex and Tom head off to stabilize the car.* As it's

---

* This is an incredibly important thing to do – if someone has a spinal injury, just a few millimetres of movement can cause paralysis.

currently on its side, we'll need to peel the roof downwards to create a large flap.

'You're inside,' I say to Sandy. I need someone in the vehicle to protect the casualties from the cutter's powerful jaws. As we slice through the roof, Sandy will use a special plastic, tear-drop-shaped shield to guard against pieces of flying metal and shattered glass.

Grumps, the generally grumpy firefighter driving our truck, confirms that the first ambulance is only five minutes away, and the second fire crew about ten. Protocol dictates that Grumps should stay by the pump, ensuring the equipment is ready and crewing the radio. However, the two children are still trapped in the back of the car with their dead father. They're my next priority. They may or may not be hurt physically, but the psychological implications for them are significant too.

'Grumps. You're with the other car. Monitor the kids and prep the car for cutting. Any signs of deterioration, I want to know.'

John is still standing with the children's mother, and I hope seeing us acting will help him instil a sense of calm in her.

I move to stand near the truck, where I can hear the radio and see the entire scene. Alex and Tom begin to cut through the roof of the hatchback and the shriek of metal struts bursting fills the air. We're making progress. I still have two injured people on the road, who were thrown from the car. They are in pain but stable – the situation is under control.

I need to establish a casualty clearing station for the injured but non-critical.

The green hatchback shifts slightly. I shout to the crew to stop and readjust the stabilization. I head over to help but am distracted by the welcome sound of sirens approaching.

The ambulance pulls up and I set aside my other plans in order to brief the two paramedics about the casualties. They leap out and rush towards me. I recognize Ged instantly. He's a robust man in his late forties and an extremely experienced paramedic. I've worked with him numerous times and on some very difficult shouts. I'm pleased to see him today.

'Gentlemen,' I begin. 'We have two cars in what appears to be an unsurvivable wreckage. One partially visible suspected fatality.' Only a medic can declare time of death, so although I can see through the window and into someone's head, it is only a 'suspected' fatality for now. I summarize the overall situation and point them towards the ditch.

'The final vehicle is there. Four passengers. A family. Father appears dead. Mother is out, walking wounded. Two children in the rear of the car, conscious and stable, but need to come out. Ged, can you assess the kids in the back of the car as a priority and then sort a casualty clearing station? I can give you someone from the second crew, but I'm working with what I have for now. The hatchback is about another ten minutes from full extrication.'

'Leave it with me,' he replies. I know they are desperate to get on with the job. The second paramedic is already unpacking the oxygen masks from their bags.

'How many ambulances are coming?' I ask.

'Two en route, but I'm calling in more. We should have a rapid-response paramedic very soon too.'

That's good news: the rapid-response paramedic can take over casualty clearing so that the ambulance crews can load up the critically injured patients.

Alex and Tom are almost ready to peel back the roof of the hatchback. I'm about to ask for an update when I notice fuel leaking from the intertwined cars further up the road. It's trickling towards the hatchback and Alex and Tom are paddling in it. This is the new priority. I can't have a stray ignition source sparking it alight.

I wave at Grumps and call him over. He raises his hand to say that he wants to stay with the children. I understand – he's emotionally involved. However, this is more important. I yell again. He recognizes my tone and doesn't hesitate this time, muttering under his breath as he jogs towards me.

'The kids are all alone and—'

I cut him off, knowing that my next orders are critical. I point to the small river of fuel glistening in the sunlight.

'Grumps. Fuel leak . . .' My expression doesn't permit challenge and Grumps understands immediately.

'Shit. Sorry, boss.'

'Get the environmental grab pack and get the powder on it.* It needs to be soaked up, and dam the drains if you see any. Keep an eye on it; make sure it works. If it doesn't, come straight back and—'

I'm interrupted by a hand on my shoulder. I turn and see Ryan, who is the watch commander of the second fire engine. He's a slight man, very tall, with fair hair. He's gentle,

---

* We carry absorbent powder for soaking up fuel spillages.

sometimes a little nervous, and friendly, but I always get the impression that I make him somewhat uncomfortable.

'Right on time,' I say. 'You're just the person I need.'

Ryan smiles. I brief him, and point towards the car in the ditch. I need him to put a sheet over the front two seats to conceal the father from view, get the roof off the car and the children out. I put him in charge of that sector and turn my attention back to the scene.

There's a mêlée of activity in front of me. It's ordered chaos. People barking directives. People hurt and crying. The mother sobbing hysterically once again; the agony of a wife who's lost her husband and a mother afraid for her children. Our day to day is assembled from other people's devastation.

I quickly survey the scene. The roof of the hatchback has been peeled back and the unconscious man is being stabilized by a paramedic. I later discover that he was suffering from a pneumothorax – a collapsed lung – and was slowly suffocating. It's a relief to know that I made the right call by prioritizing his rescue, not only with the pieces of the jigsaw I had at the time but within the wider context of the incident. According to Ged, if I had focused on releasing the children first, the man would be dead. No question.

The police arrive and I direct a PC towards a man taking photos on his phone. He is escorted behind the cordon. Grumps places plastic sheets over the deceased to preserve their dignity. I don't want their families discovering the situation from pictures on social media.

Ryan is still with the car with the trapped children – and has the roof off. A little girl, only five or six years old, with angel-white hair and clutching a small pink rabbit, is out and being looked after by Ged. He gives me a thumbs-up so I know that – at least physically – she's OK.

We have a casualty clearing station now set up, where the first-response paramedic is tending to the less serious casualties. The mother is reunited with her children. The two people jettisoned from the hatchback are side by side on stretchers and the unconscious man is being lifted into an ambulance. The car's driver will be out in a few more minutes. Then all the casualties will be in the care of the medics. We will finish by cutting the dead from their vehicles and hand the scene over to the police, who will remain on scene until a private ambulance comes to retrieve the victims. We have been here for just under an hour, although it feels like much longer.

At this incident, my decisions were primarily quick and intuitive. I had no time to weigh up every option and evaluate the optimal course of action rationally. I relied on my gut. In the previous chapter, I emphasized how important it is to consider the jigsaw in its entirety and to base decisions and actions on what *might* happen next, as well as what *is* happening now. However, it can be very difficult to find the headspace in which to do this. There are occasions when you have to deal with each piece of the jigsaw as it comes in, in turn, instinctively and intuitively, relying primarily on gut feelings.

This isn't as risky as it sounds. Gut instincts are, for the most part, intuitive decisions primed by prior experience. They are your mind compiling a jigsaw in your subconscious and then spitting out the information when you need it. I've seen dozens upon dozens of car accidents – all of them different, but variations of the same picture. In this incident, specific cues – a severe wreckage, casualties trapped in cars, a fuel leakage – activated memories or knowledge I'd obtained from previous incidents and triggered associated actions. I can analyse my actions retrospectively – weighing up various options and constructing a plan – but it isn't always possible to do this in the moment.

These 'heat of the moment' decisions fascinate me. We have little time to get them right, and if we get them wrong, people get hurt. As part of my research programme, I hoped to uncover how decisions were made in the field, to identify the way the brain processes information and therefore to find ways of improving decision-making in the field. Ultimately, I wanted to understand how we use our instinct and analytical minds when facing a disaster scene and to harness those findings into something productive.

For eight weeks, my colleague and friend Phil Butler and I lived on fire stations across the UK.* We were there six days a week, twenty-four hours a day, responding to emergencies with the fire engines and officers. This was only possible because Phil and I were ourselves serving fire officers and, while it was gruelling, it was fascinating too.

Before this, there had been very little research into fire

---

* I am eternally grateful to the participating fire and rescue services and the amazing commanders and crews who agreed to take part. Thank you to East Sussex Fire and Rescue Service, Hampshire Fire and Rescue Service, South Wales Fire and Rescue Service, Tyne and Wear Fire and Rescue Service, West Midlands Fire and Rescue Service, and West Yorkshire Fire and Rescue Service.

commander decision-making, and what existed was based on interviews given *after* an incident. The problem with this approach is that decisions can be reconstructed with the benefit of hindsight (more on this in Chapter 10). You seek ways to justify and understand your actions. You fit your responses, your decisions, post facto into a plan that was not there at the time you actioned your response.

What Phil and I did was far more immediate. We strapped video cameras to the commanders' helmets and recorded entire incidents from their perspective – so that we captured details that they might otherwise have forgotten. When we got back to the station, we would replay the recording for that commander and interview them, helping them to recall exactly what had been going through their mind at each point. The footage provided a powerful visual cue and encouraged them to remember their thoughts accurately, in the right order, at the right time.

This wasn't only useful for Phil and me, but for the commanders themselves. One described how he watched the material and saw himself the way other people saw him – which wasn't in line with the image of himself that he thought he was projecting. He told us that in his mind he was calm, measured and assertive. However, when he saw the footage, he was surprised at how shaky his voice was and how much he fidgeted. He worried afterwards that rather than instilling confidence, his demeanour might have made people more anxious. He was such a good and diligent officer that he put a proposal together for his service to introduce helmet cameras for all commanders so that they can reflect on their experiences properly. As a result, they did!

Traditionally, in the fire and rescue service, we believed

that our decisions were very analytical and followed a set sequence of gathering information, evaluating options and enacting a plan. Those decisions were very much rational, calculated ones. However, the result of the research with fire commanders across the UK showed that these thoughtful, considered decisions occurred only 20 per cent of the time. Instead, commanders were making use of their previous experiences – consciously or unconsciously – and making instinctive, gut decisions 80 per cent of the time.

Whether you use analytical pathways or intuitive pathways to make decisions is not a reflection of how good you are at making decisions under pressure. The two pathways are linked to very different parts of the brain: the analytical, rational decisions to the neo-cortex (the thinking part), and the more intuitive, associative decisions to the older (in evolutionary terms), more emotional parts of the brain, such as the limbic system. Often both pathways are viable options and your response will depend on a multitude of factors, including the situation with which you're being faced, your previous experiences, your memories, the specific pressures of the scene and how well you are coping with them.

At the road traffic collision earlier in the chapter, I relied heavily on intuitive decision-making. Equally, at other incidents – particularly those that were just as pressured but less time-critical – I've relied heavily on more analytical decision-making.

I was the strategic commander of a wide-scale flooding exercise that affected several towns. The difficulty with flooding is that, while you can pump out the water, there is often nowhere to put it. It can be a bit like trying to scoop water from a sinking boat. You empty it, but it quickly fills back up again.

I had two more long-term options for diverting the water. The first would have resulted in the flooding of a hospital. We'd have needed significant resources to evacuate the building. The second would have flooded a major electricity sub-station and would have taken out the power to over 5,000 homes. I knew that both were likely to cause significant disruption. So I needed to work out the 'least worst option'.

The stakes were high, but the situation didn't require an instant decision as the road traffic collision had done when it came to prioritizing casualties. Nothing was going to change dramatically in the short term – the town was already flooded and the situation wasn't escalating. An instant decision wouldn't save more lives. I was able to gather more information, for example, from other key agencies who could explain in more detail the implications of each option. The electricity company could tell me exactly which areas would be affected and what contingencies they could put in place. The chief executive of the hospital was able to explain the implications of closing the hospital, including the impact on vulnerable patients, the difficulties associated with moving them and the potential death toll. I was able to discuss the options and to consider the pros and cons carefully and alongside other strategic commanders.

We deliberated for around an hour, and eventually opted to divert the water away from the hospital, accepting the impact on the electricity supply. It was not ideal, and it did affect thousands of people and thousands of homes, but there were contingencies to continue providing power to key infrastructure – such as hospitals – with generators. I was confident that closing the hospital would have resulted in more fatalities.

Another example of an incident that required a substantial degree of analytical decision-making was the fire in the underground tunnels in Holborn in 2015. I was the most senior commander (the monitoring officer) at the scene. The fire took place in a twenty-five-metre section of Victorian tunnels containing high-voltage electricity cabling and gas pipes that run beneath the surface of Kingsway, in the heart of London's West End. A gas leak was burning like a jet flame in the middle of the street for thirty-six hours. The fire had a huge impact – theatres were closed, offices evacuated, several hundred people were displaced from their homes, and thousands more residences were without power. It was a technically difficult fire which showed just how complex incidents in major cities can be.

Again, it was a high-stakes incident – the fire was affecting a vast number of people and was making international news, there was gas leaking and several hundred local residents had been evacuated – but the pace was slower. The fire was static. It wasn't getting any worse, or spreading. There was time and space to make an analytical decision about the best way to deal with the fire. We needed to consider the risk posed by the cabling, the safety of the firefighters and the need to restore power to the city. We explored various options and were able to put the fire out without sending firefighters into the tunnel.

Most decision models – including the one we used to use to make decisions in the fire and rescue service – assumed all decisions were made in this way. A situation is assessed, plans are made and then actioned. Therefore, by implication, if you weren't making analytical decisions at every incident, you weren't commanding properly. Yet our research had uncovered that 80 per cent of a commander's decisions were intuitive.

After an event, decisions were being scrutinized – by colleagues or, should something have gone seriously wrong, the courts – but the decision-making model didn't reflect fairly how commanders' brains worked in a naturalistic setting. Firefighters could be made to feel as if they were doing something wrong, or badly, when the model against which they were being judged was insufficient and didn't reflect the brain pathways that they would actually use.*

I'm not suggesting that intuitive decisions are always perfect. There are times when responding directly to a piece of the jigsaw will elicit the right response for that piece, but might have a catastrophic impact on the rest of the incident. You may fall into a decision trap.

For example, you might have a fire in one room and can see the pressure building up in the movement of smoke around the windows. Techniques can vary, though you could break a window from outside and allow it to ventilate before you put crews into the building. This will improve the situation and prevent a potential backdraft. However, if you haven't considered the rest of your jigsaw, and there are firefighters already in the building, this

---

* Einstein once famously said, 'If you judge a fish by its ability to climb a tree, it will spend its whole life believing it's stupid.' That feels very pertinent here!

response might introduce air to other parts of the building, encourage the fire to burn more intensely and endanger the firefighters, or compromise their safe route out. You need to bring the firefighters out first.

Equally, an analytical decision in the face of a situation requiring an immediate response may waste precious time when lives are at stake. Our brains evolved these two pathways – analytical *and* intuitive – for a reason: to ensure humans could adapt to their environment.

The reason I looked at these two pathways throughout my research was to try to understand them and how they might make the fire ground a safer place for firefighters. I didn't simply want to describe how decisions were being made and to find the potential traps. I wanted to help commanders make better decisions. The first step towards this was developing a technique that recognized the value of intuitive decisions, but acknowledged the importance of questioning the rationale behind them, in order to avoid decision traps such as James's confirmation bias and tunnel vision at the Portcullis House fire in Chapter 3.

I wanted commanders to prompt their unconscious consciously and question how their actions link to their goals. I wanted them to open up their jigsaws and weigh the benefits against the risks.

This underpinned the development of something called the Decision Control Process. It was my aim to create a framework that acknowledged the natural brain processes we rely on and mirrored the way the brain actually works. I wanted to develop techniques that would work with our brain processes, not against them.

The Decision Control Process did just this. It recognized

both slow-time analytical decisions and intuitive ones. I knew that it wasn't realistic, or desirable, for commanders to use one or the other – brains don't work like that – but here was a technique that could be applied and would guard against decision traps when relying on gut instinct.

It set out a series of prompts that could be used consciously to reaffirm a goal, project the anticipated outcome and rationalize the benefits and the risks. In short, when a decision is made – whether analytically or intuitively – but before it's actioned, commanders would ask themselves three questions:

- Goal – what do I want this decision to achieve?
- Expectations – what do I expect to happen as a result?
- Risks versus benefits – how do the benefits outweigh the risks?

This could be done quickly, every time a decision was made. Reaffirming the goal linked the decisions to the overall aim; projecting the anticipated outcome increased situational awareness and further developed the jigsaw; and examining the benefits and risks consciously ensured better decisions overall.

My research team and I discovered that the technique worked: commanders linked their actions to their plans, joined up their jigsaws, and situational awareness increased significantly. Most importantly, the technique didn't slow down decision-making. It was practical and the feedback was great.

It also recognized intuitive decisions where previous decision models had failed to do so. This step alone helps protect commanders from unfair scrutiny, should they be in the unenviable position of reliving a fateful incident and being judged on whether they did enough. It levelled the playing field.

Following the outcome of the research and the development of the Decision Control Process, a team and I rewrote the national policy, building on the old decision model with this new information. It recognizes both intuitive and analytical pathways, and introduces decision controls to harness the best of each and improve decisions. We all benefit from this technique now – it's taught to fire commanders across the UK. It will take several years to bed in, but at least the journey has begun. It has also been included in the national doctrine for all emergency services dealing with major and complex incidents – the Joint Emergency Services Interoperability Principles (JESIP). So other blue-light colleagues are benefitting too.

In fact, the response from around the world has been beyond anything I ever imagined. I've had interest from fire and rescue services in Hong Kong, Australia, Europe, Colombia and the USA, and more recently from the military and from medics.

The amazing thing is that I very nearly didn't take the chance to build this research programme after finishing my PhD. A former senior colleague who I very much respected – and still do – couldn't see the value in the research initially. There was considerable pressure on me to drop the idea and take up the more traditional path to progress my career that I had thus far rejected. Although I knew that it would really irk him, my belief in the potential benefits of the research was greater than that fear. It was important to me and I knew it was the right thing to do. Thankfully, once the work was established and well under way, he was very supportive, and continues to be to this day.

It was a difficult decision at the time and took more courage than I thought I could muster. It turned out to be the best choice I've made in my career and it springboarded me to the once unimaginable position of being a leading expert in the field of risk-critical decision-making within the emergency services, with acclaim and research awards from all over the globe. Further evidence, if you need it, that you sometimes have to trust your gut when making decisions.

6

Judged on Shadows

'It takes many good deeds to build a good reputation,
and only one bad one to lose it.'

Benjamin Franklin

THERE ARE hundreds of versions of you. There is the 'you' that you know, but is that who your sister knows? Or who your partner knows? What about the version your boss sees? Or the barista who serves you your coffee every day, who you greet with a smile and an exchange of pleasantries? Or how about that person who met you just once, when you were having a terrible day? Just how different are the different versions of you? And how do those different versions come to be?

When you meet someone new, you probably make quick, instinctive judgements about who they might be. It's very possible that you don't even know that you're doing it. You evaluate them based on their outfit, their job, the way that they speak, the environment in which you meet – and these judgements are not necessarily accurate. The version of that person that you see is distorted by the lens of your own perception.

As a result, I think it's rare to be assessed on your true character. Abraham Lincoln famously said, 'Character is like a tree and reputation like its shadow. The shadow is what we think of it; the tree is the real thing.' More often than not, you're judged on your shadow. Someone will draw a conclusion – sometimes warranted, sometimes not; not necessarily rooted in fact, but instead in beliefs and opinions – and come to know a very specific version of you.

In a work environment, in particular, reputation is key. For some people, this is a blessing. They'd far rather be judged on their reputation than their character. They like to hide behind a veneer, the version they like other people to see. For others, it's a constant source of frustration. They don't think their reputation is fair; it's not accurate. Think of the colleague who is flying high despite having zero talent and a ridiculously sub-par output. Or the nice guy who works hard and late and never misses a deadline but who hasn't been promoted in a decade. Think about the coworker who huffs and puffs when asked to do a very simple task. Maybe it was just a bad day, but you can't help it – the version of them that you know is lazy and entitled.

Your reputation has a phenomenal impact on your day-to-day life, whether you have a good reputation *or* a bad reputation. It affects the way your colleagues initiate a conversation, the way they anticipate your response, the way they welcome your contribution – or not. These interactions then affect your response, further shaping the version of you that they see.

Martin was a station commander in his late thirties, and much younger than many of his counterparts. He was also much brighter. Neither were attributes that particularly endeared him to his colleagues. The issue was further compounded by his baby face and full head of thick brown hair. He could comfortably have passed for being in his late twenties. His features were slightly puffy and he had a round face, with greyish eyes. He was a little like how you'd imagine the stereotypical cartoon dork; the one who has grown up, filled out and made the switch to contact lenses. His gangly build didn't lend itself to packing on the muscle, but as a keen triathlete, he was certainly incredibly fit.

Martin was used to being different. The area he had grown up in was economically deprived but his family was middle class. His parents had bought a new-build executive home in an area of regeneration, and although they weren't exceptionally well-off they were comfortable. Their lives had a level of stability that others in the area could only dream of.

Martin went to school with children from the local estate and was picked on and ridiculed for being 'posh'. He grew up to be socially awkward and found it hard to interact with others. However, beneath his quiet and subdued exterior, he had an unshakeable grit, shaped by years of bullying and social isolation.

He attended university before joining the fire and rescue service, his first venture into the world of work. His family had wanted him to opt for a more white-collar profession, but Martin couldn't be dissuaded. He had a strong sense of social justice and his parents' disapproval only served as a powerful motivator. At the age of twenty-two, he was accepted on to a recruit course and became a trainee firefighter.

Martin was a quiet, reflective man. He was very conscientious and liked to plan things meticulously, whether it was his schedule for the week or the ingredients for a meal. He didn't cope very well with any deviation to any aspect of his plans – which was sort of at odds with the spontaneous nature of a career in the emergency services.

As part of my research programme with Cardiff University, I profiled the personalities of over 100 firefighters across the UK.* I wanted to see if there was any correlation

---

* The research tool used was the Revised NEO Personality Inventory (NEO PI-R). Firefighters were asked questions about their thoughts, feelings and goals, and their responses were analysed to determine their personality type across five main areas: openness to experience, conscientiousness, extraversion, agreeableness and neuroticism. For further information on the NEO PI-R, see Bibliography, p. 276.

between specific personality traits – like Martin's – and the way individuals exhibiting those traits made decisions. We didn't find a direct link, but I did find that firefighters are typically extroverts who are supremely confident in their abilities and very spirited too. Martin, however, was different. He was introverted, thoughtful and happy in his own company.

As firefighters, so much of our work involves teamwork. Cooperation and collaboration are key. When Martin joined the service he was awkward. He struggled to find a way to bond with the other firefighters on his watch, and as a result he quickly developed a negative reputation. He was often ignored. When someone mentioned him, everyone else would roll their eyes. He was never praised, only criticized, and for anything that was in any way less than perfect. His middle name was Richard and so he was nicknamed 'the dick'. He wasn't treated as part of the team and his colleagues weren't particularly supportive.

He felt isolated and often did things by himself, even tasks that typically require input from others. The results were mixed. He needed the opportunity to learn from his peers and to work alongside them to produce better results, but he wasn't getting that experience. So his reputation developed until he was considered not only anti-social but also useless. His reputation fed expectations and those expectations fed his reputation. I worked with Martin a couple of times and witnessed just how damaging this cycle can show itself to be.

I glance down at my watch. The digital display is blinking four zeros; it's midnight.

While most of the world is asleep, I've responded to a call for a derailed freight train carrying a load that included a quantity of hazardous chemicals. Each of the twelve carriages is transporting at least fifty large containers, and some have tipped over on to the tracks. Access is difficult – the closest entry point is about half a kilometre away – and we are conveying most of our equipment and crews on a special road-to-rail vehicle.

I'm tired and irritable. At this time Gabriella is just a baby and I haven't slept properly for days. I know I'm not at my best. It's a cool night, late in September, and I'm grateful for the full moon illuminating the scene of operations. The dozen carriages lie concertinaed, dew evaporating from beneath them. The unmistakeably bitter smell of burned motor oil shrouds the scene. The wreckage lies snarled up on the tracks like a huge metallic snake.

The train driver is trapped. Her cab is badly mangled and her legs are crushed under the dashboard. The crumpled metal is blocking our access.

The incident itself is complex, but my plan is relatively simple. My priority is to get the driver out safely. I then need to get the chemicals cleaned up and the containers moved to a safe place. Ideally, I want to be able to open the adjacent tracks by 6 a.m. when the commuter trains are due to begin running. However, I'm not thinking about that in too much detail just yet. The driver's life is in danger and she comes first.

Three crews are working to cut her free from the wreckage. It's hard work. The cutting gear weighs around 10 kilograms – imagine holding a beagle above your head for an extended period of time – and it's physically challenging as well as emotionally draining. Every firefighter is pushing themselves, ignoring the ache in their arms and the sweat dripping down their temples.

The other crew members are busy removing the containers. Fortunately, there appears to be relatively little chemical spillage. Martin is already on scene when I arrive and I appoint him operations commander. He's my eyes and ears at the scene. Martin is bright, attentive and more than capable of figuring out the right tactics to release the trapped driver and remove the risk of the chemicals simultaneously. I believe in him. This is his chance to shine.

Commanders traditionally inhabit a relatively macho world; there is a constant battle to be the alpha. The best way to achieve this is to be the hero, the lynchpin, the leader of the pack, and one way to make yourself look – or feel – like the alpha is to diminish the efforts of others. Humans are naturally competitive and while we might not be using spears any more, we've all witnessed displays of dominance and attempts at positioning. We've all experienced the nasty co-worker who belittles others as a result of their own insecurity.

You do not need to blow out other people's candles to make yours burn brighter, but many do, and that sets the tone for the culture in which we work, day in, day out. Martin has had his candle blown out a few times too many and it's getting harder and harder to relight.

I really feel for him. I know that he struggled working alongside a team in a watch setting when he first joined the service – he simply couldn't command the respect of his peers – and so he decided that the only way was up. He was bright and able to progress through the ranks relatively quickly. He chose positions in specialist departments, which took him away from the station-based environment that had made him unhappy for years. He was excelling and eventually reached the rank of senior officer.

He worked for me briefly – as one of my station commanders – and although he came with a terrible reputation, I was keen to see where his strengths lay. Truthfully, he did struggle to interact with others, especially in a group situation and even more so when others were vying to be in charge, but he had a clear head and strong sense of morality. Despite all of the mud thrown at him I never once heard him run a colleague down. Even when it was thoroughly deserved.

With seniority comes the responsibility of commanding more serious incidents. However, a good manager – who might excel in regular, day-to-day tasks – doesn't always make an effective commander. And the most effective commanders don't always make great managers. They may have the requisite training and experience to tackle a crisis brilliantly, but their management style might not be so effective in other situations.

However, I think it's tricky for an officer to adopt entirely different management styles in different settings. I don't think that you can be one person in the office and a different, more autocratic version of yourself on the incident ground. Every encounter with a colleague – wherever it takes place – forms part of your shadow. It informs their view of you. The one thing that will always reach an incident faster than you is your reputation. People anticipate the way you will respond, your behaviour, your way of working. Are they expecting a positive, helpful leader who instils confidence, or a stressed dictator who confuses authority with abusing their power?

When Martin's career took him into non-operational posts, he didn't work on scene at incidents for a number of years. He knew what he was meant to do, but you can't put a fire out with bullet points from a list you know by heart. You need to bring it to life with experience. Martin needed to build the emotional and sensory memory banks that would form his mental models. The detail in the sketch. He needed a little help, some support while he found his feet, the opportunity to get a little more experience. I hoped that the position of operations commander at the derailed freight train incident would offer that opportunity.

I've established a command room in a small maintenance yard about half a kilometre from the scene, and I've based myself there. It is our point of entry and exit, and offers a clear view of the tracks. I can coordinate the activity at the scene as well as communicate with the police, the railway operators' officer and the paramedics, all of whose cooperation is crucial.

I have a small group of officers in my command team. Where possible, I match a person's skills and experience to the role that needs doing, and sometimes the areas they need to improve or gain experience in, as I've done with Martin. The roles aren't predetermined – they change at every incident – and so does my team. Sometimes I get the A-Team, made up of the most experienced, able officers.

At other times, I have to pull a premiership performance together from Vanamara National League players.

The team in the command room are scribbling information on to the whiteboards, updating the locations of the fire engines and specialist appliances as they arrive and are manoeuvred on to the scene. The board to my left outlines the main risks: the danger posed by the containers of chemicals, the lack of light, the hazardous tracks underfoot and the toxic materials that might seep from the mangled carriages. I work through the new details, updating my jigsaw.

I turn to the radio operator and ask him to contact Martin for a situation report. I'm expecting a clear, confident answer detailing what's happening at the moment, the ongoing strategy, the potential hazards and the predicted resource requirements. I want to know how the extrication of the train driver is progressing and how many carriages have been cleared of spilled chemical containers. These milestones will give me a sense of our progress, and confirm whether or not the operation is proceeding as expected.

I peruse the boards as I wait for the update. The command room is unusually quiet. It is just me, my command support officer and a radio operative. All of my officers are briefed and deployed, and I've completed the multi-agency briefing. I have time to catch my breath and evaluate our progress. However, the moment of relaxation dampens the adrenaline in my system and I begin to feel tired again. I am trying to stifle an enormous yawn when my logistics officer, Vinny, walks in.

'Keeping you up, are we, boss?' He chuckles.

'Everything's keeping me up, Vin,' I reply. 'I've a baby trained in torture tactics. Sleep deprivation's her speciality.'

Vinny smiles. 'They're all like that, I'm afraid.'

'Tell me it gets better.'

Vinny frowns. His expression is that of a plumber about to give an extortionate quote.

'Get used to it,' he says. 'The sleepless nights don't stop. Just wait till she's a teenager!'

I roll my eyes. 'How's it going on the ground? I've just asked for an update from Martin.'

He raises his eyebrows and grimaces, his expression somewhere between wry amusement and dismay.

'That good?' I ask.

He sighs and I know he's going to be dismissive of the efforts at the scene.

'The thing is, boss . . .'

The radio starts to crackle with the update I've requested. I raise my hand, pointing to the radio, and Vinny stops talking. I recognize Martin's unmistakeable nasal tone. I strain to hear the message. The reception is poor, despite the extra measures we've taken to boost the signal. I gather that crews are making progress, but that the driver has not yet been released. Martin confirms that one carriage is clear of chemical containers. However, according to the information on my boards, we've cleared the chemicals in *five* carriages. How have we gone backwards from five to just one?

'Vinny, I thought we'd cleared five? What's happened?'

Vinny looks at me and shrugs. 'What do you expect when you put "the dick" in charge down there?' he says.

I glare at him and he throws his hands up in defence.

'All right. All right. My bad. I'm sorry. It's your call, boss. But, with respect, he's not up to it.'

I hate when someone prefaces what they're about to say with 'with respect'. It normally means they're about to be anything but respectful, and Vinny's unwelcome attack on Martin is frustrating. We are at an impasse. Martin is a bright and capable officer. He has an unfortunate reputation for poor performance on the scene of operations, but I don't think it's warranted. Martin is being judged on his shadow.

At this point, it doesn't really matter what Martin does or doesn't do. Vinny will focus on something that confirms his own perspective, which corroborates his belief that Martin is incompetent. Anything that goes wrong – a delay, an injury, a problem – will be as a result of Martin. Vinny is revealing his own confirmation bias, much like James demonstrated in Chapter 3.

'That's enough, Vin,' I snap. 'We'll pick this up later.'

Vinny is arrogant: he thinks I'm referring to Martin's performance, rather than his own inappropriate conduct. I like Vinny, but it's simply unacceptable to run another colleague down in circumstances such as this. I have to bite my tongue – if I weren't so tired, I'd be able to control my emotions better – but I know firing off will only make the situation worse. I'll bring it up at the debrief.

Vinny knows I'm not happy and offers to go to the scene of operations to check on the situation. I'm not convinced I'd get an objective response so I decline his offer. The command room is still quiet, so I take advantage of the lull in activity and go myself instead. It will help to update my jigsaw and a blast of cool air will quell the fatigue.

The tracks are heavily lit and cramped. There is very little space between the rails and the shrub-covered embankment. I catch myself on a thorn as I scramble to reach the front part of the train. Our equipment is strewn across the ground, and Martin is briefing the crew who will enter the train and assist with cutting. It is cold and damp, but beads of sweat are trickling down Martin's cheeks. The others, by contrast, look incredibly relaxed – a little too relaxed.

Martin shuffles nervously, visibly surprised to see me. I ask for a briefing and he explains that there were five carriages containing vats of chemicals. One has been cleared by our crews. A paramedic reported that there were no casualties in five of the carriages, a message which got distorted on its way back to the command room. Martin's message reflected the true situation. As he speaks, he shifts his weight from side to side and he struggles to maintain eye contact for more than a couple of seconds. He thinks that he's cocked up. I don't think he has.

He is repeating negative patterns of behaviour, however – allowing his mental model for his performance to undermine his capabilities – and I need to interrupt that cycle. I take Martin aside and tell him I think he's doing a good job. He has a good plan. His situational awareness is sound. He smiles. His shoulders relax. I speak loudly, making sure that others can hear me. My presence encourages everyone to sharpen up. They know that I'm watching and professional pride means no one wants to be picked up in the debrief. You know how computer screens flick quickly from online shopping back to emails when a manager walks the floor? It's just like that.

Martin stands taller. He calls one of his officers over and suggests an innovative method for extricating the driver using pneumatic rams to push the crumpled dashboard away from her legs. It's an adaptation of something called a 'dashboard roll', which we use when we have a similar entrapment in a car. The officer nods. He understands what Martin is suggesting; he likes the idea, he thinks it will work. It does.

Under Martin's command, the crew rescued the driver. She survived. Her leg needed amputating, but she lived. I hope that Martin took something away from the experience too. I hope he understood that his contribution was not only important but essential.

Confidence breeds confidence and panic breeds panic. I have seen mediocre performances delivered with such incredible confidence that they almost deserved a round of applause from the fire ground. Confidence on its own isn't enough – it's a shadow – but it's very hard to fight fire without it. If you're about to enter a burning building, you want the orders to come from someone who understands the risks, who knows what they're doing and who has a clear vision for the incident response. However, if your commander's demeanour, their body language and their attitude scream 'I'm not sure', that anxiety is contagious. You will begin to feel nervous, a little more stressed, and that will affect how you perceive the situation and process rapidly changing information. Their lack of confidence affects your performance. Their reactions make you think the situation is more severe than it is. You start to look for signs that confirm it – confirmation bias – and so you notice things that make the situation look worse. You wonder if the risks they're asking you to take aren't worth the benefits. If they're not sure, how can you be? Your stress levels increase and the cycle continues.

One of firefighters' greatest strengths is their ability to work together. The power of a tightly bonded team can overcome seemingly impossible problems. However, their greatest strength can sometimes also be their greatest weakness. Within a crew, or any tight-knit group, there will often be an alpha personality – not necessarily the formal leader – whose perspective tends to be the dominant one, and deviating from it is seen as being disloyal to the group. The bond is so tight that group cohesion is maintained above all else. This can stifle challenge to the popular perspective and this affects how individuals are perceived within a group. If the alpha thinks you're useless, there will be little opportunity for you to make a constructive challenge. Everyone thinks you're useless.

For officers like Martin, who don't quite conform to their peer-group norms when it comes to personality traits, it can therefore be very difficult to find your place within a tightly knit team. It isn't simply a result of being socially awkward either. It's group dynamics. In some instances, I've seen people ousted from the security of a group (and not just in the workplace) for being too ugly, too handsome, too good at something or not good enough.

It is essential that individuals feel supported in their work environment, not only from a well-being perspective, but in terms of performance too. If they don't feel psychologically safe within a team, they won't contribute. They won't share their ideas or insights, their hunch on where a heated cylinder might be lurking or where the fire might spread.

People fear being punished or humiliated for speaking up. This belief can be reinforced again and again – often accidentally – by so much as a badly timed sigh or a glance at the clock. As a commander, you might have a team of bright, energetic, engaged firefighters, and while they might be perfectly confident facing fire, they are scared to contribute their thoughts to the response effort. This means you're not getting the best out of your team. If team members are anxious, nervous, demotivated or afraid to speak up, the commander has failed. The same applies to any boss in any work environment.

One industry that realized the importance of contributions from the wider team is aviation. The voice recordings from the cockpits of several planes involved in tragic air disasters revealed that crew members had tried to warn the captain about something critical. However, because of the power dynamics, the information was often not clearly conveyed. Planes crashed. People died.

Following these disasters, a team of NASA researchers explored the extent of the problem. John Lauber, one of the team and a research psychologist, told a reporter about an occasion when he had entered a cockpit prior to the captain's arrival.

The flight engineer said, 'I suppose you've been in a cockpit before.'

'Well, yes,' John responded.

'But you may not be aware that I'm the captain's sexual advisor.'

'Well, no, I didn't know that,' said John.

'Yeah, because whenever I speak up, he says, "If I want your fucking advice, I'll ask for it." '*

The industry realized that a pilot's absolute position of unchallenged power meant that other crew members felt they couldn't point out mistakes or flag concerns. The researchers suggested an innovative approach called Crew Resource Management (CRM), which aimed to reduce human error by focusing on the way people think and respond. (Sound familiar?) A key objective was to foster a culture where authority could be respectfully questioned. CRM is now a global standard in aviation.

Being in charge is more than simply telling people what to do. The way others respond to you can make or break an incident. You need to be able to engage your colleagues; to inspire and encourage them to see your plan through. As a leader, you also have a responsibility to take care of people in your charge. And you need to be brave. Being brave doesn't mean not being afraid of something. Being brave means doing something despite being afraid. Sometimes that means taking risks to ensure that we not only play to our strengths but also develop new ones. It's hard and none of us gets it completely right all of the time. We all make mistakes, but I hope that every firefighter on my fire ground feels respected, supported and valued, whether they're well liked or not.

I know from personal experience how it feels to be an outsider. When I joined the fire and rescue service at eighteen, the previous few years had been difficult ones. When I left home at fifteen, support from the authorities was

---

* www.vanityfair.com/news/business/2014/10/air-france-flight-447-crash

woefully inadequate.* At sixteen, hungry and tired from sleeping rough, I visited the office of my local social services, hoping to find my old social worker.

However, substantial cutbacks and huge caseloads had taken their toll, and I was met with a relatively empty office. I was redirected to the council and denied a priority place on a housing list because I was already homeless. Priority is instead given to those who are *about* to become homeless. (I can't resent that policy; finding yourself homeless is horrible and I wouldn't wish it on anyone.) I was sent to a housing association and added to the bottom of a list. I never heard from them again. I was too young to access benefits and there was no one to guide me through the complicated, bureaucratic housing system. I was on my own.

So I learned to take care of myself. I lived hand to mouth, selling as many copies of the *Big Issue* as I could. However, I simply couldn't make enough money that way. So I left the crowded city of Newport – where several sellers were competing for enough sales to feed themselves for the day – and took my magazines to a little town called Monmouth, where I was the only one selling them.

I stayed there from six in the morning until seven at night every day. I slept rough some nights, found shelter in

---

* Things have improved somewhat since the Children Act 2004 was passed, which insists on inter-agency cooperation in matters relating to a child's well-being. Any agency aware of the maltreatment of a child must now make their findings known to other agencies that might be able to assist in protecting a child who would otherwise go unmonitored.

derelict buildings for others. Other alternatives for a bed for a night came with a price I wasn't willing to pay. I once slept in an underpass and woke up to find a very intoxicated man urinating on my sleeping bag. The scrawny stray dog that had befriended me jumped out and sought vengeance for the both of us. I tried hostels, but they were often full and, to be honest, I didn't feel particularly safe in them.

I eventually saved enough money to put down a deposit on a *very* cheap rented flat and moved to a town in the South Wales valleys, near a retained fire station that I hoped one day to join. I wanted to help rescue other people because no one had rescued me. The context was different but the sentiment was exactly the same. When I mentioned my plan to people, they laughed and pointed out how small I was. It would be impossible, they said, which only made me more determined. At every knock-back or low point, I'd picture their sniggering faces and it spurred me on.

I found some employment that provided a simple yet sustainable life. By the time I was eighteen, I was working in a factory making ready meals and had at least three bar jobs at any one time. Then I set to work on chasing my dream: becoming a firefighter.

An optician told me that my eyesight was too poor, that it would never be good enough for me to join the fire and rescue service. I almost cried, but then I got angry instead. I was absolutely not going to give up. I went to the library and found a strange book on eye exercises. I did them diligently, eight times a day, for months. They didn't work. At all. In fact they just gave me a migraine. I needed a better plan. I went back to the library and found a book on memory techniques. At my next eye test, I distracted the optician briefly – I'd also found a more general book about eyes in the library – and memorized the bottom two lines of the board, just in case. I passed the eye test.*

I took the rest of the physical and written tests, and then had my interview. It all seemed to go well but it was hard to tell for sure. I was scared that I hadn't done enough, but all I could do was wait for the post to arrive. Every day, I ran to the door and grabbed the mail from the postie, like a child waiting for a letter from Santa. Eventually, it came. In an unassuming brown envelope bearing a red stamp and the fire and rescue service emblem. I read my name four or five times. I sat down at a small wooden table in the living room and I ripped it open.

'Dear Miss Cohen, I am delighted to inform you that you have been successful . . .'

I blinked and read it again. Then I leapt up and started screaming, bouncing around the room while my little black mongrel dog, Menace (the same stray that had befriended me on the street), jumped and barked beside me.

You create your own opportunities. Your background doesn't determine where you end up, only where you start. A brilliant reputation – or great connections, or a top education – may grease the wheels, but your ability to succeed is directly correlated to your grit and the effort you put in to something. I refuse to believe anything less.

---

* The stringent standards for eyesight set for entry into the fire and rescue service have since been revised and perfect vision is no longer a requirement.

I often wonder whether qualities such as tenacity and determination are products of nature or nurture. I know it's a very complicated question and I think most of us agree that the answer is a combination of the two. The way you respond to a given situation is determined by your personality – and some of those traits will be inherited – but your response is also influenced by previous experiences, your values and your sense of what is right. All these things can be learned from your family and friends.

When I think about my own family's history, I believe there's some evidence of grit further up the tree. My Jewish family had been rooted in Morocco for generations, with branches in Rabat and in Oujda, on the Moroccan–Algerian border. My grandparents had a comfortable and happy life until my grandmother was assaulted with a machete in a pogrom in Jereda on 8 June 1948. Her head was hacked open by an attacker in a frenzied anti-Jewish massacre that saw at least forty-three Jews slaughtered and more than a hundred injured because of their faith.* Her first born, my uncle Gadi, who was just three years old, was hiding beneath her skirt as she was left to bleed to death on her own doorstep. My grandfather was elsewhere, braving the violent streets in an attempt to find a police officer to come and protect his family, whom he'd left barricaded in the house. He narrowly avoided being hacked to death

* The records state forty-three people were killed, but according to my grandparents the number was much higher. They estimate at least sixty died. The records, particularly at that time, weren't always reliable either – my grandmother doesn't have a birth certificate, for example, so she simply guesses her age.

himself: his knife-wielding assailant was shot by the police as he leapt on my grandfather from behind.

When my grandfather returned home much later, such was the severity of the violence, he discovered their eldest child, thankfully, unscathed. He was told that my grandmother's body had been taken to a makeshift morgue, which he went in search of as he needed to see her. He found a shack, guarded by a single police officer, and bribed his way inside. He searched through a pile of mutilated bodies – many were people he knew – and found his wife at the bottom. With incredible effort, he managed to drag her out. As he did so, she gasped for breath. She was alive. Injured, but alive.

It seems my grandmother was grittier than most. She had been bleeding profusely, severely injured by an extremist with a machete, and yet – even surrounded by death on all sides – she had managed to cling on. My grandfather's tenacity and will to protect his family – and find his wife – shows that grit came from his side of the family too. (Indeed, Mike's grandfather escaped from Bergen-Belsen concentration camp – twice. I hope there's plenty of tenacity coursing through Gabriella's veins.)

My grandparents fled Morocco soon after. Europe wasn't an option, and Jewish people from North African countries often struggled to get visas elsewhere. So, after many generations of living as dhimmis* and experiencing

* An historical term – meaning 'protected person' – for a non-Muslim living in an Islamic country and granted special status and safety under Islamic law in return for paying a tax. Dhimmis had fewer legal and social rights than Muslims, but by paying this tax they were permitted freedom of religion and worship.

growing persecution in their native land, they went where they thought they'd be safe: Israel. And they were right. They lived a simple, happy life there. Subsequent generations of their family would be free to practise their religion without fear for their safety. My grandfather made shoes and my grandmother made a home. They raised five children in a two-bedroom flat. The first of their children born in Israel was my father.

Maybe my grandmother's fierce determination was passed on to me through our DNA. Or maybe it was something she learned from her parents and I learned it from her by copying her work ethic, her tenacity, her refusal to give up. I saw it many a time throughout my early years when she lived with us and helped to nurse my father. I also saw it in my father when he refused to give up. Given six months to live, he fought on for nearly six years.

When I joined the fire and rescue service, nearly two decades ago, I was the first woman to serve in that particular station. I'd expected that. What I hadn't expected was the overwhelming sexism I faced at the time. I wasn't exactly welcomed with open arms. I remember a number of surprising conversations that started with 'I don't agree with women in the job. No offence to you, but I just don't agree with it.'

I found myself apologizing for my own existence. It didn't take long for my teeth to show and my teenage wilfulness to kick in, though. I enjoyed casually responding with, 'I know what you mean. I feel the same way about morons in the job. There's no way we should employ them but then, here we are! No offence to you, of course.'

The all-male crew at that particular station refused to talk to me at fires and took great pleasure in watching me struggle. One incident involved a burning car that had rolled down an embankment. I was left to clamber across branches with a hose, completely on my own, to extinguish the fire. The three other members of my crew – including the officer charged with our care – stood behind me in a small semi-circle. They didn't even watch to make sure I was safe – or effective. I glanced back occasionally and saw them smirking.

They didn't call me by my name. They called me 'split arse'.* I was eighteen at the time and some of those men had daughters a similar age. I tried so hard not to let it get to me. I wanted to be liked, yes, but more than that I wanted to be a firefighter. It was ironic, given their passion for a tight-knit firefighter 'family', that they were so ready to ostracize me – an act that goes against what it means to be a firefighter. Their behaviour fuelled me to do better, to be better, and to make damn sure I never made anyone else feel like that.

It wasn't just name-calling. During the early years of my career, other firefighters messed with my kit and blamed me – with deadly seriousness – for the scent of a rotting kipper left in a departing colleague's locker as a joke. I've been sexually harassed more times than I care to remember: I've received unsolicited 'dick pics' and been told things *have* to be different for me because I'm not a bloke. I didn't call it out at the time. I was young, insecure, and I

---

* If it's not a term you're familiar with, google it – but not from a work computer, perhaps. I refuse to give it any air time here.

believed the balance of power was against me. I was afraid
that speaking out would affect my reputation so I put up
with it. I now deeply regret doing nothing, and I've worked
hard to put it right and always call it out now.

However, I must add a caveat: these bad experiences
made up only a small part of my first posting.

Thankfully, so much has changed in the last two decades in terms of culture, genuine inclusion, and in recognizing the contribution made by everyone, not just
those who fit the stereotype. I joined at the cusp of a
change. There are people who came before me and fought
for the rights within the service that I take for granted
today – simple things, such as being able to have long hair,
and bigger things, such as having fire kit designed specifically for women, with protection in the right places for our
bodies. As the result of a lot of hard work, determination
and personal sacrifice from people with great vision –
some who didn't fit the mould, and some who did but were
willing to use their power and influence to change things –
I can put my hand on my heart and say with great confidence that I do not recognize the fire and rescue service
from the year 2001 in the one I work for today.

It's a world apart. It's brilliant. And that's why I want
more women to join and make their mark. Not because I
think we need to fill an arbitrary quota, but because we
need to choose the next generation of firefighters from the
best of the best. At the moment, we're only choosing from
the best of the best of the small group who are attracted to
the stereotype of what it is to be a firefighter, which doesn't
necessarily reflect the broad range of work we do today.
We need to widen that pool.

To do this, I do think we need to challenge the fringes of acceptability and call out everyday sexism. I've had to block people on Twitter because I receive pure bile as a result of regularly contesting the use of the term 'fireman'.* Not from serving firefighters, I might add. It's mainly from those outside of our industry, those who have such a sense of privilege that, to them, the pursuit of equality feels like oppression. We need to continue to challenge inequality, even if it's hard and even if we're shouted down.

I have a daughter. I know I can't wave a magic wand and undo centuries of accepted social norms and unconscious bias, but when she grows up, and she comes to me complaining about an injustice or inequality (and I know she – like all of our daughters – will still face them), I refuse to simply say, 'I know how you feel.' I hope to always say, 'I know how you feel, but here's what I did ... And this is what you (and everyone else) can do going forwards.'

I'm also so heartened by the wave of support from men – particularly through the HeForShe movement – who recognize the need for change for the sake of their daughters, sisters and partners. I see more and more guys in work calling things out now, and fighting for real inclusion. Gentlemen, thank you. We appreciate you.

---

* The term 'fireman' was replaced with 'firefighter' in UK fire and rescue service employment contracts during the 1980s. Given the thirty-year career span at the time (since increased to forty years), there are very few – if any – serving firefighters that have ever actually been 'firemen'.

I soon moved to a different station where I worked alongside some amazing officers, truly inspirational people who became like older brothers to me. I look back on those years – and people – with huge fondness and I wouldn't change a thing. During my career, I've made some of the best friends of my life, had people stand up and support me in the face of some of the worst moments, learned I can do things that I never thought I was capable of, laughed so hard I thought I was going to give myself a hernia, found people who I dearly, dearly love, and had the privilege of working with people whom I trust – wholeheartedly – with my life.

However, knowing how it feels to be mistreated – to have an undeserved, negative reputation within a group into which you are desperate to be accepted – had a significant impact on me and has influenced how I treat people today. I choose my words as carefully as I choose my actions, and I never underestimate the bearing that you can have on another human being. I don't always get it right, but I'm self-aware.

It has also moulded the way I behave as a commander. I will not see anyone belittled or treated badly in my team. I will not accept the excuse of 'command style'. I don't care if you're under pressure. I cannot tolerate people barking orders and assuming that, because they've shouted them, those orders will then be carried out.

It's crucial to consider the subtleties of your message, even when the situation is intense. Are you confident that you have articulated your meaning accurately and that the message has been properly received? How can you make risk-critical decisions and see your strategy enacted if your requests are delivered in an aggressive, potentially unclear manner? Everyone's feeling the heat: it is not acceptable to use 'command and control' as an excuse to behave poorly towards people. Not only because it's unkind and cruel, but because it also affects performance negatively.

I've never been great with authority figures, and commanders who behave in an autocratic manner have done nothing to help my view. I've seen so many people in trusted positions get it wrong – from parents, to teachers, to bosses. My secondary school missed so many signs of my vulnerability. One teacher even saw me selling the *Big Issue* and, when I called out to him, he looked nervously at his shoes and walked straight past without saying a word. I was fifteen at the time, in the middle of my GCSEs, and I remember bursting into tears because I knew then that no one – actually no one – cared at all. As an adult, I can recognize that my teacher just didn't know how to deal with the problem. But the ostrich who buries his head in the sand shares only his arse with the world.

I'll never forget the time one very senior and very drunk officer told me that I wasn't going to – and shouldn't – get promoted because I 'didn't have a cock'. My temper got the better of me and I snapped that I might not have one but I was clearly working for one, which I considered to be the same handicap. I'm not encouraging this sort of response (although it did make me feel much better at the time) but I find it really hard to hold my tongue, even for the betterment of my career – when I disagree fervently with what someone is saying.

I'm sorry to say this – perhaps you won't want to hear it – but being an adult does not make you right. I knew this as a kid (and didn't I get the almightiest whacks for it!) and I know it now. As for being the most senior individual in the room, or at an incident? It doesn't make you right either. When I first took a rank, a very wise ally told me not to forget the four other people in the crew. He said that although I make the call, they come up with the ideas. However, if it all goes wrong – they will be four witnesses. There is strength in working as a team, but call it wisely!

The first time I established a major incident room, I thought I knew exactly what I needed to do. However, an incredibly experienced watch commander – who technically was significantly more junior than me but had worked in that kind of environment many times before – challenged my interpretation of the situation. We were using a new building and had to set up the space to mimic our normal working environment. His suggestions were absolute gold! Whereas I was ready to make do, he recognized instantly – from experience – that it wasn't going to work. He managed to access a locked side-room which then doubled up as a separate 'command room'. It allowed me space away from the other separate cells, or functions, to pull all the streams of information together. Because he felt able to speak up and contribute, without feeling as if he was putting my nose out of joint, we quickly established a much more effective room than I would have been able to do alone.

I think it's wise to have a healthy suspicion of authority figures (ironic, I know, given my current position) but the truth is that I want to be challenged. I want my officers to be able to say to me, 'I think it might be better if we do this.' Or, 'Have you thought about this?' Or, 'I disagree with that course of action.' If my decision is the right one – if I can rationalize it with logical, valid points – then we're no worse off. If my call turns out not to be the best option, we have time to change our position. I've seen others, who are feared, fail spectacularly as a result of their pride. I don't want that to be me.

7

Mental Preparation

'Train hard, fight easy.'
Alexander Suvorov

A YOUNG MOTHER clutches her child to her chest in the middle of the street. She is screaming; a wail that wavers between shrill and guttural. Tears are streaming down her face, spotting her pastel-blue shirt. She is holding the baby so tightly that I fear she might crush it. The infant is swaddled in a pink-spotted blanket, its face hidden from view.

An enormous, muscular police officer dressed entirely in black points an automatic weapon at the woman and her baby. He is a model of power and dominance; she is one of desperation. He wears a balaclava, a helmet and goggles conceal his eyes. He is unidentifiable except that – even through all his body armour – he carries a very strong and distinctive scent, with notes of sandalwood peppering the air. There is a surreal discord between this faceless behemoth and his human smell.

He shouts at the woman to get down on the floor, to put her hands behind her head. She moves slowly towards him, trembling, pleading for help. A white muslin cloth hangs loose, dragging in the dirt and splashing mud on to her faded jeans.

This woman, dressed in a summery shirt and leather jacket, might be exactly what she seems to be: a terrified civilian. Or she could be another terrorist, another pawn in an elaborate plot, an extremist hiding an explosive device behind an infant, hoping to reach the cordon and detonate it amongst the first responders.

As the mother of a young child myself, I'm instinctively drawn to her. Her fear is inscribed on her face and I can only imagine what she must be feeling: disorientation, panic, that heady mix of adrenaline and terror.

What would I do?

I'm silently begging, pleading with her to get down on the floor because, for fuck's sake, he's actually going to shoot you if you don't. If I were standing where she is, though, with a terrorist running rampage in the shopping centre behind me and a masked police officer in front, both armed with automatic weapons, what would I *really* do?

Five consecutive gunshots reverberate around the street, immediately followed by more screams. The vibrations rip through the air, echoing in my bones. The woman drops to the ground, wailing, squeezing her baby against her. Everyone ducks, covering our heads with our hands, spinning in position to find the source of the shots.

'Clear!' an armed police officer shouts.

The bullets came from the police. Their sniper spotted an opportunity and fired through a ground-floor window. He's taken out the terrorist.

The 'hot zone' is now clear and specialist crews can enter and tend to the injured.

My radio crackles as firefighters and paramedics weighed down by ballistic protection rush into the building, carrying lightweight plastic stretchers and medi-kits.* The building is now a 'warm zone' † and our objective is to extricate the wounded.

---

* Firefighters wear burgundy bullet-proof protection – helmets and body armour – over a burgundy boiler suit when responding to terror incidents.

† The 'warm zone' is the area where non-police responders with appropriate personal protective equipment (such as ballistic protection) can deploy to assist casualties and to extinguish fire.

I follow two firefighters who are carrying a stretcher. Their boots pound the floor under the weight of their gear and they breathe heavily under the physical strain. The inside of the building is eerily quiet, devoid of the usual bustle of the Saturday crowds. The shop fronts are peppered with bullet holes. Blood is smeared across the walls where people have tried to escape the unfolding terror. There's a fire a few floors up, set by the gunman as he whipped through the building, shooting as many civilians as possible.

Survivors call out to alert the crew to their positions.

One of the firefighters I'm behind runs straight towards the voices and into a stationery shop. The other turns and puts his hand over his mouth, trying desperately not to retch. He inhales deeply, then turns and enters the shop too. I pause and peer around the door. The stands have collapsed and multicoloured folders are scattered over the floor, covered in blood. A woman is lying in the corner, speckled with gunshots and clearly dead. A second woman is on the floor, bleeding and unresponsive. The shallow rise and fall of her chest is just visible beneath her baggy, blood-soaked T-shirt. A man is writhing in agony, his trousers soaked with blood, a red stump protruding from the torn fabric. His dismembered limb, blown off by the force of an improvised explosive device, lies several feet away.

'He fucking shot us,' he shouts at the firefighters. Then, 'My wife . . .' He points at the unresponsive woman. 'Is she OK?'

The firefighters have one stretcher and, between them, the strength to carry one casualty from this building. The bleeding woman has the least chance of survival. Should they prioritize the man?

There's no textbook answer. It's a judgement call. So they make an intuitive decision, taking cues from each other, a simple glance or a nod of the head. One makes straight for the injured woman and puts pressure on her wounds. The second puts a tourniquet around the man's leg.

'We're going to get your wife out of here,' he says. 'And then we'll be back for you.'

'Please – don't let her die. We've got three kids . . .' His voice trembles and then trails off. He's realizing that their lives have been altered irreversibly.

Together, the firefighters strap the woman to the stretcher and head back towards the entrance.

As they brush past me I overhear their brief, whispered conversation.

'Was that the right thing to do?' murmurs the first.

The second shrugs. 'There is no right thing, mate. Not when you've got nutjobs running around with Kalashnikovs.'

He has a point.

After 9/11, the emergency services across the world altered their state of preparedness to protect people from terrorism. Attacks were no longer the realm of the unexpected. Everything changed again after Mumbai when, in November 2008, ten gunmen targeted civilians in public areas of the Indian city – railway stations, cafés, hospitals and hotels – over the course of four days. There were hundreds of casualties and 164 people died.

The Nag Hammadi cathedral massacre took place in Egypt in January 2010. Coptic Christian worshippers leaving the midnight Christmas mass were targeted by gunmen in a shooting that claimed nine lives and injured eleven others.

The world then watched as eighteen people were taken hostage in a siege in a coffee shop in Sydney in 2014. Two people were killed. We looked on from in front of our television screens as tourists were targeted at the Tunisian beach resort of El Kantaoui in 2015. Thirty-eight innocent people were murdered. Closer to home and in the same year, the offices of *Charlie Hebdo* were attacked in Paris and twelve people were killed. The heinous attack at the Bataclan theatre also followed in November that year and claimed a further 130 civilian lives. Then we saw nine innocent worshippers shot dead by a white supremacist in the Charleston Church shootings in America. In all of these attacks, guns were the weapon of choice.

Our world has become a very dangerous place to call home. The threat from terrorism has been at an all-time high for the past few years and we've seen attacks such as the murder of Fusilier Lee Rigby, Manchester Arena, Westminster Bridge, Borough Market and Finsbury Park all on home soil. Yet despite this, I firmly believe that people in the UK are among the best protected in the world. The emergency services are incredibly well trained and, should the invisible safety net that works so hard to protect us somehow allow someone dangerous to slip through, we are ready and trained to respond. At major incidents all emergency services work together. As well as extinguishing fires, firefighters are often responsible for evacuating casualties in such situations and ensuring that they receive medical attention as quickly as possible.

It's therefore imperative that firefighters train alongside other blue-light responders. Knowing how we operate on our own simply isn't good enough. The members of an orchestra can learn their part of a symphony individually, locked in separate rooms, but no one would then expect the full ensemble to play together perfectly on the first night of the Proms. That would be ridiculous.

The same is true for the emergency services and, in the context of a major incident, the stakes are so much higher. Training together builds our mental models and creates a shared situational awareness. We know that responding to a major incident requires the effort of a large number of people, several different teams and multiple agencies. We know too that major incidents are dynamic environments. There are so many different micro-operations moving in tandem that, very quickly, individuals may find themselves responding within the confines of their own experience, based on their own mental models and their own situational awareness. Essentially, individual responses can become detached and separated from the overall strategy, like an out-of-tune violin disrupting the sound of an orchestra. This can happen unintentionally but it can make difficult decisions even more challenging.

When the fire and rescue service trains with the other emergency services, we learn how the other agencies think, the roles they play and how they respond. We then know not only our own song sheet, but theirs too, which helps us to better predict what might happen in certain situations and the effect our decisions may have on other parts of the jigsaw. Most importantly, training together builds personal relationships so that, should there be a major incident, we're working alongside people and services that we know and trust.

You've probably noticed by now that I introduce both real-life incidents and training exercises in a very similar way. This is because it's absolutely essential for training scenarios to feel realistic and for participants to feel the same levels of stress, pressure and uncertainty that they would do in real-world situations. There is a fear that burns inside you when someone is screaming in pain, desperate for your help, that changes the way you process information. A lifeless, faceless casualty dummy matched to the weight of an adult (although admittedly useful for honing rescue technique and increasing strength) simply doesn't prepare you for responding to, caring for and saving the life of a real human being. Neither does it prepare you for helplessly watching them die, despite your best efforts.

The exercise described at the beginning of this chapter was one of the most realistic I've ever attended. It was specifically designed to prepare emergency services personnel for a terror attack. The terrified woman with the baby was an actress. The man with the dismembered leg was an amputee, trained to react in an authentic way to medical treatment (both correct and incorrect techniques). The other casualties were actors who had given up their time voluntarily to assist us. Very often they have experienced traumas of their own in the past, which makes their contribution all the more phenomenal.

I've been fortunate to experience many excellent training exercises, but I've also experienced some terrible ones.

I remember arriving at one particular scenario that was so unreal it bordered on a Monty Python sketch. I turned up at a training yard as the incident commander to discover a tall, narrow drill tower, which was meant to be a ship. An old, battle-weary officer was hanging out of the third-floor window, shouting gibberish, pretending to be the ship's captain who spoke no English. I tried so hard to take it seriously, but I couldn't help giggling. At which point the officer reached into his top pocket, pulled out a corned beef sandwich and threw it at me.* I ducked and it hit my pump operator square on the nose.

Although incredibly funny, the lack of realism was distracting. Even when the crew finally managed to focus, we each found ourselves imagining different types of ships, and before long the exercise fell to pieces. Opportunities to train are so important that we can't afford to waste time on activities that don't reflect our actual working environment.

Imagine you're an Olympic swimmer. You train in a huge pool three times a day. You're strong and you're fast. In terms of distance, you could swim the English Channel twice over – no problems whatsoever. However, if you decided to dive into the Channel one morning, you might find that the physical skills you've been finessing in the familiar training environment of the pool are useless in the context of a choppy sea. You're putting the same skill to use in many regards – it's just swimming – but the context has changed. You've been working hard at your

---

* I genuinely have no idea why he chose to keep his lunch in there. Perhaps he had a penchant for sweaty sandwiches.

speed and how to turn efficiently, but what you need in the Channel is strength and a different technique.

The same reasoning can be applied to developing mental skills; different situations require different techniques. There is little use honing your situational awareness, decision-making, communication, resilience or leadership skills in a scenario you will never find yourself in.

You have likely experienced this discord between preparation for an event and the reality of it before. Think about how you might put in the work for a job interview or an important presentation. You practise at work and at home, locked in a room on your own, rehearsing answers or your speech until you are word perfect and feel completely confident. Then the big day arrives but, despite your careful preparation, you fall to pieces.

Last summer, I was preparing to give a TEDx Talk. I had to speak for eighteen minutes, uninterrupted and unsupported by cards, slides or prompts. Essentially, I had to learn a six-page monologue. And I needed it to appear unpractised and natural. (A little like having to spend three hours in front of the mirror to achieve that perfect 'just got out of bed' look.)

I practised diligently for several hours a day, for a number of weeks. I'd rehearse in the car, in the bath, in front of the mirror, to the dog and in front of any friends and family willing to hear my talk again. I was word perfect. However, when I finally stood up before the crowd my mind went blank on at least three occasions. I knew the entire transcript inside out, but I had only prepared in safe environments – never under pressure. Let's be honest, the dog was never going to troll me on Twitter if I was crap! I hope that I managed to pull off the blank moments as pregnant pauses, but I was acutely aware of the silences, even if the audience wasn't.

The missing key was context. You need to practise in conditions as close to those you'll eventually find yourself in. For an open-water swimmer, it's the sea. For someone giving a talk as I did, it's in front of people who make you feel at least a little bit nervous (or at least a harsher critic than the dog). And for emergency responders, it's under immense pressure . . . and not from the fear of having a corned beef sarnie flung at you.

It's therefore incredibly important for the fire and rescue service to replicate the intensity and pressure of real incidents in our training exercises. We need to be sure that commanders are used to experiencing stress and are trained to overcome it, because stress not only affects our ability to operate in emergency situations – by reducing our ability to make effective decisions – but also has a significant impact on our emotions.

For example, stress often induces feelings of anger or irritability, which, in turn, can make communication more challenging. Without clear communication, your team might not understand your jigsaw, your perspective, or your intentions. Your negatively perceived attitude elicits negative responses in others. Panic spreads. The relationships within your team break down and the team members lose confidence in you. You aren't being an effective leader any more.

I once knew an officer called Gerald. He worked in the office down the corridor from me and, both being early birds, our paths often crossed in the morning. He came across as calm, rational, balanced; a really lovely man. He seemed totally unshakeable. Until, that is, he stepped on to the incident ground. He found the stress so overwhelming that he would flap and panic as soon as he pulled on his boots. He would snap and bark at people because he found it difficult to explain what he wanted them to do, and because of his demeanour he had a terrible operational reputation. He was often labelled at best an idiot and at worst a bully.

One day I met him for coffee a short walk away from our headquarters. We were outside of our normal working environment so I decided to have a frank conversation with him. I asked if he realized how people perceived him on the fire ground. He admitted that he didn't enjoy operations. And 'ops' is the thing you're supposed to love as a fire officer. It's a huge part of what we do; it becomes part of our identity. Gerald had found himself trapped in a difficult cycle: he didn't like operations, so he didn't train, which made him feel unprepared, and so he grew to dislike operations even more.

This is why practising and preparing under realistic pressures is so important. For many years the fire and rescue service relied very heavily on real-life experiential learning for commanders and firefighters. They honed their skills over the course of their career, becoming more and more adept at responding while experiencing extreme stress, and became better commanders and firefighters as a result. However, over the past decade, the number of incidents that the service attends has reduced by 50 per cent. We respond to a much greater range of incidents, but we have a thinner experience base of each type. Today's commanders therefore receive only half the operational experience of their predecessors. However, fires don't burn any less fiercely these days, so we need to train twice as hard in order to ensure our officers are adequately prepared.

The other huge benefit of realistic training exercises is that our commanders and firefighters can become more experienced earlier in their careers, and our entire service is stronger as a result.

Like Gerald, there have been times in my career when I've wished I had more training. I once attended a very difficult incident where a young horse had become trapped in a ditch. The horse had been panicking and writhing around, and the walls of the ditch had collapsed. At that time in my career, my only experience of large-animal rescue was from a twelve-page policy document that focused primarily on the hazards and risks associated with large animals (mainly covering the facts that they were a: large, and b: animals). The bullet points offered nothing to help me with the situation I was facing. I remember feeling woefully frustrated.

We eventually rescued the animal but, unfortunately, she'd broken her leg and had to be put to sleep. Watching that magnificent creature be euthanized was one of the saddest events I've ever witnessed at an incident. The vet could sense my frustration and assured me that the break likely occurred when she fell, but I couldn't help wondering what might have happened had we been better prepared and rescued her sooner.

This incident was just another example of the 'what ifs' that crystallized my intent to develop research that can help improve the way we make decisions. It is one thing to uncover the qualities needed to make decisions successfully in the emergency services, but I wanted to work out how we could use that knowledge to ensure that all firefighters and commanders have opportunities to hone and rehearse the necessary skills.

The research explored three very different training methods: virtual reality, traditional training exercises and 'live burns'.* My team and I developed a range of different scenarios, each crafted around ten key points where a decision would be required. Some were decisions about how to save the lives of others, some related to unexpected turns of events, and some to moral conundrums. The scenarios included a house fire with a child trapped inside, a collision between two cars with critically injured passengers, and a fire in a shop.† Eighty-four incident commanders from across the UK volunteered for the research, and I worked alongside some brilliant crews who really brought the scenarios to life.‡

---

* It's not as cruel as it sounds, I promise!
† For more information on the different scenarios, see Notes, pp. 265–7.
‡ Commanders volunteered from Avon Fire and Rescue Service, Buckinghamshire Fire and Rescue Service, East Sussex Fire and Rescue Service, Greater Manchester Fire and Rescue Service, Hampshire Fire and Rescue Service, Hereford and Worcester Fire and Rescue Service, Isle of Wight Fire and Rescue Service, Merseyside Fire and Rescue Service, Northamptonshire Fire and Rescue Service, South Wales Fire and Rescue Service, Tyne and Wear Fire and Rescue Service, West Midlands Fire Service, West Yorkshire Fire and Rescue Service, Wiltshire Fire and Rescue Service, and the Fire Service College.

We experimented with virtual reality (VR) in the training suite at the Fire Service College. Some commanders fully embraced it, others were less enthused. One commander was adamant that a 'computer game' wouldn't be realistic enough to teach him anything. Ten minutes later, he was stepping over an empty space – a cone on the screen – and then fighting with a disobedient avatar. We had a convert. However, I'd drastically underestimated the physical side-effects of virtual reality. Several commanders in the first cohort became very queasy; I had to warn the second group not to drink the night before! The other downside was that commanders were working on their own – there were no real crews to direct, just the avatars on screen. Virtual reality is incredibly immersive but it is also highly simulated and therefore the least realistic of the three training methods.

The second method was standard drill-ground training, which involved the most traditional exercises used to prepare commanders and firefighters. Incidents were run in special facilities where we could simulate fire and smoke. We used the same range of scenarios that we did with VR, painstakingly recreating them and even constructing a section of motorway with scrap cars to simulate the road-traffic collision. This phase was more realistic – commanders could direct real crews who responded in real time – but it was glaringly obvious that the experience still lacked realism. Despite our best efforts it was impossible to create the same level of uncertainty as crews would likely find at real incidents.

The final method I wanted to explore therefore needed to be acutely realistic. I wanted to match the uncertainty and the unpredictability of a real fire. I needed to introduce a real sense of risk. I discovered that Hampshire Fire and Rescue Service actively seek out and rent sites with derelict buildings for months, and sometimes even years, where they subsequently set fire to the disused buildings before demolition takes place. They innovatively and proactively use these opportunities to secure real-life experience for their firefighters and commanders – known as 'live burns'. It was perfect. With their permission, I crafted a selection of abandoned buildings into the different scenarios we'd developed and then set fire to them. All of them.

Commanders and crews responded to the live-burn fires as though they were real ones – because, of course, they were. To ramp up the tension, inside one of the buildings I set a fire-proof box containing a recording of a small child screaming and crying.* Some commanders were horrified, thinking we'd actually left a child actor inside a burning building.† Their anxiety revealed their elevated stress levels.

In each environment and with each training method, I trained half of the incident commanders to use the Decision Control Process introduced in Chapter 5, and asked half to operate as they normally would. The results were striking. For commanders operating as they normally would, the reliance on intuitive, gut decisions was evident regardless of environment. Just as we found at real-life

---

* With thanks to my colleague who kindly recorded his child's bedtime tantrum!
† I think there might have been some ethical concerns had we attempted that . . .

incidents. This showed us that the training environments are an effective comparator to real-life incidents. There was some suggestion that the more realistic the context, the more intuitive decision-making took place. It could be that when you are feeling stressed – and perhaps anxious too – you are more likely to rely on your instincts, although we need to do more work to know for sure. This supported our earlier findings – detailed in Chapter 5 – which highlighted the importance of preparing commanders for making intuitive decisions as well as analytical ones.

We also noticed that when commanders used the decision-control techniques, they gave more explicit consideration to their operational goals, always ensuring that each and every decision contributed to the overall aim of the exercise. They also anticipated the consequences of their decisions more often and with more accuracy.

It was possible to measure their levels of situational awareness by analysing the words they were using during the exercise and the interview that followed the scenario, searching for any indication of whether their jigsaw was limited to the here and now, or whether they were thinking ahead. I was looking for evidence they were predicting the likely outcomes of actions, or the likely development of the situation.

Those commanders using the decision-control techniques achieved the highest level of situational awareness five times more regularly than the commanders who weren't using the technique. Critically, these benefits didn't slow down their decision-making.

These results were a first for the emergency services. Never before had we taken a really hard, evidence-based look at how we train. We already knew the importance of a good dose of stressful, emotional realism and how it could stretch commanders and prepare them for the unforgiving pressures of the incident ground. However, we also learned that command training simulations were effective in engendering similar decision-making processes to those we had observed in real life. Your brain responds to decisions in the same way. We realized that training simulations could be an effective part of our general strategy to build experience in the fire and rescue service.

As well as on-the-ground firefighters, the fire and rescue service also works with many commanders from multiple agencies who coordinate the gravest, most challenging of incidents remotely. Our most senior strategic commanders – those who attend the most complex incidents – typically have incredibly busy and challenging jobs. It is therefore often difficult for them to find the time to undertake drawn-out, in-depth, intensive training. So each opportunity where they can do so must be maximized.

The training exercise detailed at the beginning of this chapter involved a variety of first responders, including police officers, paramedics and firefighters. A senior strategic commander from each of the emergency services was involved too and these individuals made up the Strategic Coordinating Group (SCG). As a team, they were responsible for making decisions that would determine how the entire emergency services responded to the incident.

Jonathan Crego's decision-trap exercises – such as the one with the explosion in the tunnel in Chapter 2 – are designed to encourage commanders to rehearse the skills that they will need during that once-in-a-career incident.

However, he can't sit down with everyone over a bottle of wine, as he's been known to do with me. So he developed a sophisticated system of computer systems and scenarios – the Hydra system – that commanders use to rehearse those skills. What's more, he makes this cutting-edge technology freely available to all emergency services across the UK. He runs the Hydra Foundation as a non-profit resource.

He wanted to apply his training model to multi-agency decision-making, so that it was possible to rehearse the skills needed for making decisions in a team as well as individually. He wanted to focus on how agencies train senior commanders with this level of responsibility and, in particular, those who typically manage challenging incidents remotely. I affectionately refer to this programme as 'Armageddon training'.

A series of computers is set up in different rooms and linked to one central computer – the 'brain' – that is assembled in the main training room. A large group of strategic commanders – all of whom have experience in, or may soon make up, an SCG – are briefed together. An incident is described – a fire in a chemical factory with a toxic smoke plume affecting several counties, for example. The commanders are then split into small groups of five or six and sent into separate rooms. In each room, there is more information on the incident, several video clips and a list of issues for which the commanders need to agree a strategy.

Some tasks might be fairly straightforward, such as listing their priorities. However, some might be more complex, such as coordinating a response to the toxic plume, including which areas to evacuate, which hospitals to shut and where to put the evacuees. Once the commanders have discussed, debated and agreed their response – as they would do in an SCG – they type it into a decision log on the computer screen.

At set points, all of the groups are brought back into the main room and all of the decision logs are displayed on the central 'brain'. A facilitator will encourage the groups to unpick each decision. Often, the groups will each have developed an entirely different approach. Sometimes one will have made a decision as a result of reasoned debate. Other times it's groupthink, where the need to maintain harmony leads to people simply agreeing with the loudest voice. Some participants are brilliant at detaching themselves from the final decision, some take challenge as a personal slight. You can learn a lot about someone's leadership style from their response, but it's a great place to learn.

Armageddon training offers commanders the opportunity to evaluate their decisions retrospectively and, most importantly, to think about how they made them and why. It's an environment that encourages everyone to recognize their natural biases. Are they someone who typically speaks over others? Are they likely to nod and agree in order to avoid confrontation? As a result of the training, they will hopefully question their response and contribution at real incidents in light of what they know about themselves. I trained as a strategic commander using this system and found it very effective. I learned to navigate difficult conversations, challenge egos and balance wildly differing perspectives.

*

In 2016, London Fire Brigade launched Exercise Unified Response, the largest and most complicated multi-agency major incident exercise ever run in the UK and backed by over €1 million of European Union funding. My team at Cardiff University was invited, along with a number of other brilliant academics, to evaluate the exercise and gather and collate the data in order to progress our research. It was the first opportunity to examine how commanders responded to Armageddon training when it involved realistic, real-time scenarios, rather than verbal briefs, pre-prepared documents and staged videos.

The exercise simulated a collapse in London's Waterloo station. A tunnel network was created in a disused power station in Kent and eight tube carriages were buried beneath hundreds of tonnes of rubble.* Over 2,000 live casualty actors were involved in the four-day exercise and there was even a life-sized dummy that could simulate giving birth. There was a system in place that allowed commanders to activate the EU Civil Protection Mechanism, which summons resources from other countries. Crews from Italy, Hungary and Cyprus arrived in the middle of the exercise to get involved and work alongside our own emergency services.

The exercise ran in real time, both for on-scene responders and for the SCG. The SCG was responsible for the on-scene activity, and also for collating information from the Mass Fatalities Coordination Group, led by a coroner, and the Humanitarian Assistance Steering Group, which

---

* The focus on realism was so meticulous that they used wallpaper mimicking the familiar tiles of Underground stations across London.

ensures that casualties receive the best possible care. The primary priority for the SCG was saving lives. However, they were also expected to focus on regaining a sense of normality, considering legal, financial and environmental implications, and the overall impact the incident might have on infrastructure, businesses, welfare and the community.

The exercise began with emergency services being called to our makeshift London Waterloo station following reports of a partial station collapse. The situation was quickly declared a major incident. Emergency services spent the first day securing the site and rescuing as many people as possible from the rubble. They established a plan to dig deeper in order to release people from buried carriages.

The second day saw a wider focus as responders tried to piece together a path to normality. Local authorities brought in specialists to provide immediate practical and emotional support to those involved, and to ensure ongoing assistance and welfare arrangements. The number of deceased was so great that it warranted the activation of the London Mass Fatality Plan. This was led by an on-site coroner, who had the grim responsibility of coordinating information on those we could not help.

The scale of the challenge presented by the buried train carriages required additional support. Reinforcements – urban search and rescue specialist teams from Italy, Hungary and Cyprus – were drafted in.* Days three and

---

* Rumour had it that one of the units housing the Italians' rescue equipment had a built-in barista coffee machine. Given it would likely be deployed for days at a time, who could deny them a coffee?

four focused on the gruelling task of digging up the final, deeply buried carriages, rescuing any remaining survivors and recovering the bodies of those who weren't so fortunate.

The Exercise Unified Response was hugely important in many ways. With Professor Rob Honey and our PhD student Byron Wilkinson, I evaluated the performance of the strategic commanders. We spent most of our time at the SCG, collecting information and painstakingly logging decisions, sequences and footage so that we could unpick and analyse the events later. The opportunity was incredible: everything was happening in real time and we were observing decisions in the context of the wider incident, rather than parachuting players into a pretend exercise. Not only did it provide a vast amount of data, but it also challenged all the emergency services to work together and taught us lessons that might otherwise have only been learned in a painful public inquiry.

I have always maintained that the work of our research team at Cardiff University should be rooted in improving outcomes for firefighters and those who trust us to rescue them. The work we do has to have practical applications. We compared our findings to the results of other Hydra training sessions we had run separately. We discovered striking similarities between the way that commanders were making a decision in the highly simulated Hydra exercises and the more realistic, real-time scenario of Exercise Unified Response. Hydra was clearly an effective method for stimulating the same decision-making processes that commanders experience in a more naturalistic setting.

However, for the first time, our research team also recognized that despite having the same principles to work from, there was a huge amount of variation in the way that groups would make decisions, and a relatively limited evaluation of alternative options – a potential decision trap. Through the Joint Emergency Services Interoperability Principles, emergency responders now have new decision controls that can be used when making a decision as a group (rather than as an individual). They are aimed at tackling this problem, helping commanders to ensure contingencies are more consistently explored and the rationale for a given course of action routinely assessed before it is commissioned.

Commanders are first encouraged to challenge the joint understanding and position of the group. Then, they're encouraged to reflect, as individuals, on whether the decisions of the group accurately mirror their own professional judgement and experiences. If the decisions are sound, they will stand up to this rigour.

Other people can benefit from these techniques too, from those in the boardrooms of companies charged with making multi-million-pound decisions that affect thousands of jobs, to families sitting down together to decide on a holiday destination. The approach helps to weed out the tacit influence of the person with the strongest position (so that you don't end up saying yes just to please your boss or your grumpiest relative!).

The decision controls are now embedded in our national guidance for multi-agency response to major and complex incidents, and taught to all strategic commanders throughout their training (in the Joint Emergency Services Interoperability Principles doctrine). Every commander from every emergency service now carries a small card in their uniform pocket that lists some key prompts, including those decision controls. It's an important reminder that decision-making is a learned skill; leadership, decision-making, situational awareness and communication aren't instinctive abilities, they're acquired with a bit of practice and a lot of hard work.

Why are we doing this?	What are the goals linking to this decision?
	What is the rationale, and is that jointly agreed?
	Does it support working together, saving lives and reducing harm?
What do we think will happen?	What is the likely outcome of the action; in particular what is the impact on the objective and other activities?
	How will the incident change as a result of these actions, what outcomes do we expect?
In light of these considerations, is the benefit proportional to the risk?	Do the benefits of proposed actions justify the risks that would be accepted?
Do we have a common understanding and position on:	The situation, its likely consequences and potential outcomes?
	The available information, critical uncertainties and key assumptions?
	Terminology and measures being used by all those involved in the response?
	Individual agency working practices related to a joint response?
	Conclusions drawn and communications made?
As an individual:	Is the collective decision in line with my professional judgement and experience?
	Are we (as individuals and as a team) content that this decision is the best practice solution?

8

The Price of Being Human

*'It's not the person refusing to let go of the past, but the past
refusing to let go of the person.'*

Anon

TRAUMA IS not uncommon. Most of us have some experience of it, be it mild or more severe. It isn't an overstatement to suggest that at least a third of us will experience some significant trauma throughout the course of our lifetime.* For me, and perhaps for those of you who've also been in life-threatening situations, my experience of trauma has served to highlight that life can be very unpredictable.

It is absolutely normal for a trauma to affect the way you think and feel, particularly in the weeks that immediately follow the event. Distressing or shocking incidents have the power to change your perception of the world and, understandably, that can be very unsettling. However, for some, a life-threatening experience will have a more serious impact on their mental health. It can cause long-term effects, such as psychological and physical symptoms, that are not only unsettling but also incredibly frightening and destabilizing.

Emergency responders are repeatedly exposed to trauma. It's our day to day. So perhaps it's unsurprising that we suffer disproportionately from poor mental health. This theory was supported by research from Mind, the mental health charity, which revealed that 90 per cent of blue-light responders have concerns about their mental health. The research also noted that emergency workers are twice as likely as the general workforce to identify the cause of a mental health concern as an experience in the workplace.†

Most of us who've worked on the front line – whether in

* Mental Health and Wellbeing in England Adult Psychiatric Morbidity Survey (2014).
† Mind Blue Light Research Programme (2016).

the emergency services, medicine or the military – see loss
as an inevitable part of our working lives. We have memories of particular incidents that stay with us, or situations
that affect us and our outlook on life. Sometimes we can
manage by ourselves, or with the help of friends and family, but sometimes we can't.

I'm standing on the side of the road, the wind biting at my
face. It is bitterly cold. Both sides of the motorway have been
closed. It is eerily quiet without the usual drone of traffic. It's a
silence I've experienced before, but one that always feels alien.

Twenty minutes ago, a small lorry ploughed through the
barrier into oncoming traffic. The lorry now stands on the
other side of the road, isolated and facing the wrong direction. The driver suffered only minor injuries and the vehicle
is empty. However, it collided with a car, which is now surrounded by firefighters, its bonnet smashed into the
footwell. The silver side panels are crumpled like tin foil.
The tarmac is peppered with detritus from the wreckage;
plastic trim and pieces of metal are strewn everywhere.

Crews call to one another as they lift the crumpled roof
from the main body of the car, shards of glass falling like
droplets of rain, and lay it safely on the ground. A young
woman in her early thirties is in the passenger seat. She is
finally accessible and officers begin to extricate her carefully from the vehicle. I move towards the car, hoping to
anticipate any problems before they arise.

The young woman is writhing in pain as a paramedic
tries desperately to insert a line into her collapsing veins.
Her long, red hair is matted with congealed blood. She has
a large gash on her head; I'd guess she hit the dashboard

on impact. As far as I can tell, there is no passenger airbag. She has an open fracture at her left elbow and white shards of broken bone protrude through her ragged skin. A trickle of clear fluid seeps from her ear, indicating a possible skull fracture. Her condition is serious and she needs surgical intervention urgently. This is the sort of care that only a hospital can provide; the emergency services are limited by what we can do at the side of the road. Incredibly, she's been conscious throughout and even dialled 999 initially.

There is no one in the driver's seat. I step closer and am overwhelmed by the sickeningly dry, metallic smell of blood. It's the same odour that hits you when you walk into the butcher's. It's sweet, but bitter, and hits the back of your throat before your nose. I peer into the back seat. There are fragments of bloodied bone and a few pink-red blobs, like a dropped blancmange. There is a torn flap of skin on one of the seats. It was part of someone's face, with an ear and a few blackish curls at the edge of a hairline. Scrunched in the footwell is a sandy beach towel with the insignia 'Mr & Mrs'. They are newlyweds returning from their honeymoon.

I look back towards the motorway. The wife has a direct, very graphic view of her husband's crumpled and mutilated corpse lying motionless in the middle of the road. Only part of his head is intact and it is badly disfigured. His remaining eye is still wide open with fear. I quickly cover him with a sheet.

I later learn that the lorry driver had fallen asleep at the wheel. He had crossed the motorway barrier and hit the oncoming car head on. The driver's side had absorbed most of the impact and the husband was killed instantly.

∗

This incident stayed with me.

I used that stretch of motorway regularly – it was part of my route home – and yet I began to avoid it. Every time I found myself on it, or a piece of road that looked similar, I would visualize the woman's final image of her husband. I'd then imagine the moment they first met, a classic Hollywood love story, and try to picture him before the broken face and flaps of skin. Then I would start thinking about her again and feel utterly devastated trying to imagine how it felt to have your hopes, dreams and plans demolished in an instant.

I even felt for the lorry driver. Perhaps it was complacency to drive when he was that tired, or perhaps it was pressure from his bosses and having a family to feed that pushed him to take a needless risk. Either way, I knew I'd never forget his quivering lip as he looked at the scene and understood that, from that moment onwards, life would never be the same again. I couldn't stop seeing his face.

Mike and I were about to get married. We were busy planning our wedding day. However, I'd suddenly find myself feeling anxious for no discernible reason. We'd be walking hand in hand or giggling at something daft and I'd think about how lucky we were, and then I'd feel overwhelming sadness for that newlywed couple.

With hindsight, it's clear why this incident affected me so negatively, when hundreds of others hadn't. I was planning a future with someone I loved and I had just witnessed a trauma involving two people at a similar point in their lives. Moreover, I'd watched a woman in my position have her future stolen in a freak accident. My understanding of the concept of 'for ever' had shifted and I didn't know how to handle the new information.

For a few weeks I didn't sleep terribly well. I was irritable and grumpy, and I often felt low and slightly hopeless. I cried when I thought no one was looking and, if caught, tried to blame it on a sad book. (I cry at *Finding Nemo* so this was a very believable cover story.) I wouldn't admit it, but deep down I felt like my plans for the future didn't matter any more, that there wasn't much point. That incident changed how I saw the world and I found myself grieving for the perspective I had before.

Mike noticed that I seemed a little down and, to cheer me up, surprised me one day with a DVD and a giant bar of chocolate.* He'd chosen what he thought was a soppy, romantic number. Have you seen the film *PS I Love You*? It certainly sounds romantic but it's all about a woman who loses her husband – the love of her life – to a brain tumour (which instantly reminded me of my father's illness). He knows he's going to die and so leaves her a series of letters to ease her grief and encourage her to move on when the time comes.

I bawled. Then Mike and I laughed at my tears and at the film choice, and then I started to open up about the way I'd been feeling. It was a huge relief to talk to someone finally, and it encouraged me to start piecing together what had happened and why I had been feeling so low.

I had felt afraid of speaking up. I didn't want people to think I was weak or couldn't cope with the inevitable trauma that we're exposed to in emergency situations. However, talking helped.

---

* We hadn't got married at this point. I know now that the sweet gestures come to an end!

Before long, my negative feelings eased and I came to terms with my recalibrated view of the world. I had learned that life isn't guaranteed, but I had also discovered that I could live with that realization. The turning point for me was hearing, just a few weeks later, that the woman in the accident had died too. I felt curiously relieved that she wouldn't have to live the rest of her life with that terrible image haunting her. So it stopped haunting me too. I think I still would have come to terms with the situation had I not received this news, but it might have taken a little longer to process.

Did I respond to this incident with anxiety? Yes. Was my response abnormal? I don't think so. The feelings I experienced were short-lived – and, I would argue, rational – and after a brief period of time they disappeared completely. Most of the people who work on the front line of emergency services will have similar moments in their careers and I think it's fair to say that, once in a while, we all cry when we think no one is watching. However, processing that emotion and finding a path back to normality is absolutely essential.

At the beginning of this chapter I mentioned that around a third of the population will experience a traumatic event in their lifetime. Approximately one in five adults will experience symptoms of post-traumatic stress disorder (PTSD) and one in twenty will actually develop PTSD.* This condition is characterized by intense and

---

* Mental Health and Wellbeing in England Adult Psychiatric Morbidity Survey (2014).

debilitating emotions, hypervigilance, flashbacks, night-mares and sometimes physical reactions, such as nausea, sweating and heart palpitations. I experienced very minor symptoms – some intense emotions, the odd intrusive thought – but for PTSD sufferers the symptoms can be extreme. It's a very real issue for the emergency services. PTSD can have a substantial impact on the lives of emergency service personnel, our well-being and our families, who inevitably have to pick up the pieces of our shattered selves.

I know an experienced firefighter, Ron, who suffered badly from PTSD after a particularly nasty fire. I hadn't seen him for some time, but I bumped into him one day outside the station where he worked. He was smoking a cigarette and looked exhausted. I liked him a lot and had worked with him previously. His facial expression worried me; it was out of character and I felt sure that something was wrong. I sensed that a few pleasantries on the doorstep wouldn't suffice and so went inside to get us both some coffee.

'Here you go,' I said as I passed him a mug. He didn't respond. He hadn't heard me, even though I was right beside him.

'Ron . . .' I repeated, this time catching his eye. 'Your coffee?'

'Thanks, mate. Just what I needed.' He took the mug and briefly closed his eyes. 'That's better,' he mumbled.

He lifted his hand and noticed his fingers, spreading them wide in front of his face.

'Look at that,' he said, slowly turning his hand over. 'They're white as ice. I've never felt them so cold; it's been like this since the diagnosis.'

'Diagnosis?' I asked.

'PTSD,' he said. 'It was quite a relief, to be honest. I thought I was going mad.'

One of the unexpected physical side-effects of PTSD is cold hands and feet. Your body is in a heightened state of vigilance constantly, responding to the 'fight or flight' hormones. It thinks that you're in danger and so it directs blood flow away from your extremities to protect your vital organs. There's therefore less blood travelling to your hands and feet, and so they can feel very cold, even on a warm day.

I hadn't seen Ron for nearly a year, but in that time he'd aged substantially. His once dark, floppy hair was shaved short and salted with grey. His skin was pale and mottled. He had been a giant of a man, tall and extremely muscular, but his shape had changed dramatically. He'd lost a lot of muscle and put on fat around his middle, which is another unpleasant side-effect of PTSD. The heightened levels of cortisol (the primary stress hormone) in your body restrict the uptake of proteins into muscle cells.* This makes it virtually impossible to gain or retain muscle. The cortisol also interferes with how lipids are stored and causes fat to build up around the abdomen.

'It's more common than you think, Ron. But it's good to see you back.'

'Thanks, Sab. It's been hard, really hard. People didn't get it . . . Neither did I.'

---

* The evolutionary explanation for this is that the proteins are instead converted into useable energy and used to respond to the imminent threat, either by fleeing or fighting.

I took a sip of my coffee. 'I'm here if you want to talk,' I said. 'And I'm here if you don't.'

'It started almost a year ago,' he began to explain. He'd attended a particularly harrowing fire in which an eight-month-old girl and her older brother had died.

Their parents were out for the night, the first time since their daughter had been born. The babysitter had been struggling to get the baby girl to sleep, but eventually the baby's eyes closed and she settled. The babysitter went downstairs to cook some chips for the little boy, who was playing in his room, but then the baby woke up. The sitter went upstairs to tend to her, forgetting to turn off the chip pan. She stayed upstairs for a while, shushing and trying to comfort the little girl, and while she was there the pan caught fire. There was no smoke alarm in the house so, when she eventually smelled the smoke, she rushed downstairs to tackle it herself. But it was too severe. She tried to get back upstairs to help the children but she couldn't push past the choking fog. She rushed outside and called for help.

Ron was on duty that night and his team were called to attend. He carried the lifeless little girl out of the house in his arms. He attempted – hopelessly – to bring her back using CPR. He placed her tiny, lifeless corpse in a body bag. He was the one who wiped the soot from her delicate face as he zipped it closed.

Then he went back to the station and got on with his shift.

It was late, about two o'clock in the morning. Normally, after a call at that time, people return and go straight to the dorm. This time was different. The crew all went to the mess and sat down. Someone made a pot of tea and they found comfort in not being alone. Some talked, some simply listened. Black humour is often our armour against unimaginable horrors, but Ron said there was none of that. Instead he sat there in silence, not listening, not speaking, just thinking. His throat constricted and it was painful to take a sip of tea. You've heard people describe a lump in their throat? Ron said this was a boulder.

He found himself on autopilot for the rest of the shift. He felt detached, a little numb, but he expected to feel better after a good night's sleep. It was his last shift of the week so he was heading home for a proper rest. He couldn't relax, though. His mind kept returning to the scene, to the moment when he zipped up the body bag, to the little girl's face disappearing.

Over the coming weeks, his symptoms grew worse. He experienced vivid nightmares and woke in the middle of the night, cold and sweating, the bed sheets soaked through. His body was responding physically to the emotional terror he was experiencing. He was safe in his bed, but his mind was somewhere perilous.

Ron kept questioning if he could have done anything differently, if he could have changed the outcome. If he'd turned left instead of right, he might have found her quicker. If he'd gone straight up the stairs instead of feeling along the corridor first. If only the hose line hadn't caught on the door frame, costing him precious seconds.

He found it difficult to talk about the experience. His girlfriend asked but he cut her off. His solution for dealing with the overwhelming feelings was to feel nothing at all.

He was numb. He used to train regularly and tried to distract himself by lifting weights in the gym, but it wasn't enjoyable any more. He watched his muscle gains slip away and grew even more depressed. He became isolated and withdrawn and stopped going to the gym altogether.

Then he stopped going out at all. He started drinking heavily at home. He was trying to self-medicate; he wanted to block out the intrusive thoughts and feelings. He found himself constantly on edge, and became ratty and irritable. People assumed this was a result of the alcohol, but he was actually experiencing a state of 'hyperarousal'. His body was constantly looking for threats and this was affecting his ability to concentrate and his overall mood. He snapped at people. He didn't want to talk about it because he didn't want to be seen as weak. He wasn't ready to ask for help. His peers grew uncomfortable around him and Ron found himself ostracized.

After several months of steady deterioration, Ron's flashbacks ramped up a level and he started to experience symptoms of psychosis. They were vivid: he could smell the smoke, taste the soot, hear the noise of the pump. He began to experience visual hallucinations. He would see the baby girl's face in car windows as he walked through car parks, in the bin when he took out the rubbish, on the TV when watching the news. He was frightened, and he realized he couldn't cope on his own.

When he attended a routine medical with occupational health, Ron broke down in front of the doctor. The doctor was concerned and, with a little prompting, Ron found the strength to eventually tell someone about his symptoms. This was the turning point for him. He could finally get the help that he so desperately needed and begin the long road towards recovery. He had taken nine months off work.

'This is my first week back,' he said. 'It's still not easy.'

Ron was still struggling to navigate his relationships with his colleagues because, despite the high incidence of mental health issues in the emergency services, such issues are still heavily stigmatized. Recently, 71 per cent of blue-light responders reported that they weren't encouraged to discuss mental health in their workplace. The general workforce figure was substantially lower at 45 per cent.*

The truth is that it was easier for Ron's colleagues to assume that he was attention-seeking than to face the reality of mental health issues. Because mental health concerns can be really frightening. Not only for those suffering, but for their friends, family and colleagues too.

Part of the problem is that emergency responders are regularly exposed to people in the midst of a mental health crisis, members of society at their very lowest point. We respond to people trying to take their own lives by jumping off a bridge, the effects of repeat fire-setters and the aftermath of a chemical suicide.† We see the worst and it feeds into our expectations of mental health, informing our mental models. If a colleague says that they have mental health concerns, or have been diagnosed with depression or PTSD, we jump immediately to those severe examples. However, we can't identify ourselves in these

* Mind Blue Light Research Programme (2016).
† A chemical suicide involves self-inflicted exposure to toxic gases or other chemicals in a confined space, usually a bathroom or a car.

extremes – we don't want to – and so it becomes more difficult even to acknowledge our mental health experiences, let alone ask for help at an early stage.

Another part of the problem is that emergency service personnel are 'fixers'. We see ourselves as the people who parachute in at the brink of catastrophe and walk the tightrope between life and death. We like to think of ourselves as 'protectors' and this can become a key part of our identity. It makes it very hard to admit our vulnerabilities, let alone expose these vulnerabilities to others, particularly other protectors who we fear might see us as weak.

The research by Mind found that the impact is worse for colleagues from minority ethnic groups; they are even less likely to ask for help. When you've been labelled – in terms of race, gender, or sexuality – as something 'other' throughout your career, you often try very hard not to be 'different'. I know from personal experience that I've felt obliged to push back against discrimination for most of my career. I don't like the implication that I'm employed to fulfil imaginary quotas or that I'm not here on merit.* In the past I've felt as if I've had to work twice as hard to be considered half as good.

For people who've lived this experience, labels around 'difference' have negative associations. It's not uncommon to adapt one's behaviour in order to 'fit in'. This is evident in how employees adhere to an unwritten dress code at work and how children in the playground want the latest

---

* Despite being common tabloid rhetoric, there are no, and never have been, quotas for the number of women or people from underrepresented groups in the fire and rescue service.

must-have toy. When you spend so much time trying not to be different, it's perhaps unsurprising that current support services have had a low uptake from minority ethnic staff. The research hasn't explored the implications of gender and sexuality, but I wonder if a study of female and LGBTQ responders would reveal similar results.

Organizations such as Mind and their Blue Light Programme are making huge strides in raising awareness of mental health issues within the emergency services, and there are now 'champions' on stations across the country who are breaking down the wall of stigma, stone by stone. It's a big wall, and it will take time, effort and diligence to achieve, but we will break it down.

At the moment, firefighters rely heavily on the strong bonds that exist within our teams for mental health support. We basically live together for forty-eight hours each week, eating together, training together, laughing and sometimes crying together. The camaraderie that develops is essential. The black humour nudges us back to normality. It is a relief to sit down with your colleagues after seeing the unspeakable and laugh when you notice that one of them has the mug with comedy graffiti on the base.* I firmly believe that laughter really is the best medicine.

I remember attending a memorial with two colleagues, Dan and James. We were all close to tears, dark glasses covering our watery eyes. We'd brought some lilies to the reception and, as it was a hot day, we went to put them in

---

* A chinagraph pencil is typically used to write our names on our breathing apparatus and record who's inside a building, but this is often seen as a far more amusing use.

some water. I was struggling to open the plant food sachet, so Dan chivalrously took it between his teeth and tore it open ... squirting the contents into his mouth inadvertently. We had been close to tears but now we were trying desperately to suppress our laughter. And it isn't just about trying to be positive; laughter has been shown to actively reduce stress hormone levels and trigger a release of endorphins. The camaraderie between crew members can thus be a really important part of finding light in the darkest moments.

I'm not denying that there is plenty more the fire and rescue service can do to ensure that cases of PTSD are recognized and responded to quickly and effectively. However, I also believe that there are steps we ourselves can take to prevent PTSD from developing. The service can encourage and train firefighters to be perceptive and decisive and assertive, but it can also build resilience so that our emergency responders have tools at their disposal to handle the extreme psychological stresses of the job. As I have progressed through the ranks, I have become much better at pre-empting my own responses to the emotional impact of incidents. Now, when I sense the pressure building or start to feel overwhelmed, I take five minutes away from the command unit to tour the incident ground. For me, this offers a brief period of quiet, some perspective, and a chance to refresh my jigsaw as well as my state of mind.

It is also important that officers and commanders are able to pre-empt their responses in the aftermath of an incident. After the road traffic collision at the beginning of this chapter, for example, I was disturbed as much by my own response as I was by the incident itself. I'd seen plenty of death by that point in my career, and other more harrowing scenes, so it was a shock to be so affected by that particular call. I didn't know that my response was normal; I didn't know what to expect. If I had, I think I would have been much less distressed.

Early in 2018, Professor Rob Honey and I were keen to expand our research programme into the essential skills with which to equip emergency services personnel and how best to train for the role, and so we joined a cross-disciplinary team at Cardiff University which is supervising a brilliant doctoral project that explores this exact issue.* The programme of research will train over 1,000 police officers in trauma resilience. It will then compare the resilience of the officers with the training against that of a 'control group' of officers who haven't had it. This will primarily demonstrate the effectiveness of training, but our hope is that it will also give officers the tools they need to protect themselves against the detrimental effects of repeated exposure to trauma.

This will benefit not only the individual responders, but also every person who relies on our emergency services: responders who have their mental well-being looked after are more likely to stay in the job for longer and will have more experience as a result. In theory – and should it work – it is something that could be taken up more widely.

This job has certainly had a negative impact on my state of mind. The repeated exposure to trauma is, at times, exhausting, and it has definitely had an inevitable bearing

* Led by PhD student Paula Riella.

on the way I view life, death and suffering. However, it has also affected me in a very positive way. I take the time to appreciate my life and those in it every single day. I focus regularly on enjoying the little things in life: the flowers smell a little bit sweeter in the face of a hard day, the colours look a little bit brighter because I'm alive and so are the people I love. I take nothing and no one for granted.

However, I'd be lying if I didn't admit to a sometimes overwhelming fear of loss. I try to overcome it by concentrating on what I have, and for the most part, that works. It's important to me that those around me, my family, have a positive outlook and not a fearful one. Every night, when Mike and I put Gabriella to bed, we go through the same routine. As a family, we tell each other three things that we're grateful for, things that we're really happy to have in our lives. Sometimes they're big things, like our health and somewhere to live. Sometimes they're not quite so momentous, like a new series of *Peaky Blinders* (which on Maslow's hierarchy of needs sits somewhere between wi-fi and food).

What I love most about this approach is that it takes our fears, our negatives, and it spins them into positives – if a relative falls ill we naturally feel vulnerable about health, but look, we're healthy. That's not everything, but it is something. Then we each share something that has made us happy throughout the course of the day. Gabriella tells us her stories in great detail, reliving the memories while we share in her highlights.

Finally, we consciously try to do one random act of kindness each day. This can be something small, such as offering to help someone struggling with their shopping, or something more impressive. I have to confess to being very proud of the time Gabriella asked for donations to raise money for a bicycle ambulance for a village in Malawi instead of having birthday presents. I'm never quite that selfless. Mine tend to be giving way at a junction or not letting the door swing closed on someone I can't stand. At a push, it's passing the money-off vouchers that I'm given at the tills to someone waiting in line behind me.

This job isn't easy – and I would be doing a disservice to my colleagues if I pretended that it was – but for me, it's heartwarming to see the next generation, even in just one home, finding something positive in an outlook born through the tragedies of life.

9

Brains and Behaviours

'A window to the soul.'

English proverb

I CAN SIT in a café for hours, hands wrapped around a mug of steaming coffee, watching people wander past. I like to think about them – I wonder where they're going, what's on their list of things to do, whether they're an interesting person – and they, I assume, are thinking something too. Is it positive, negative, or even very private? There's no way for me – or anyone – to know.

What if there was a way to reveal someone's secrets, though? What if there was a window into another's mind? That doesn't sound too bad, does it? It might be nice to know what your partner, your colleagues, your friends really think of you. What if this window wasn't inward-facing, though – a window you could peer through – but outward-facing, like a projector, displaying your inner-most thoughts to anyone who glanced in your direction? Suddenly it doesn't sound quite so much fun . . .

The good news is that all the nasty details are safely confined to the depths of your mind. However, while your thoughts aren't directly observable, it is possible to detect what another person is thinking or feeling. I've referred to 'micro-expressions' in previous chapters and they are fantastic clues to what's going on in the mind. However, your behaviours also reveal your thought processes. For example, the way you articulate a problem can highlight the type of decision pathway that you're using.

A friend tells me that she has a problem at work. She really needs the day off on Tuesday to accompany her sister to a hospital appointment. However, it means missing an important meeting. If she's not there, her company will likely lose out on a new contract. This is her sister, though . . .

She talks me through her options – not going with her

sister or missing the meeting. She then decides that she's going to pull in some favours and try to change the meeting to the afternoon so that she can go to the hospital in the morning.

Having listened to her work through the issue, I would be feeling pretty confident that she was using analytical brain processes. However, if she'd told me she needed Tuesday off and in the same breath said that she'd decided to pull a sickie? I'd say she was using intuitive brain processes. Her actions, her language and her behaviour has revealed something that she perhaps didn't mean to.

In my line of work, it's incredibly useful to be able to analyse someone's state of mind through the things they say and the way they say them; the way they hold their body and how they respond to events as they unfold. Each of these behaviours can reveal how well someone is processing information, how well they might be forming a mental jigsaw of the situation, how confident they are and how effectively they are making decisions.

It's a scorching August afternoon. The heat is warping the air above the tarmacked road in front of me as I drive to the rendezvous point. I park between a fire engine and another officer car. I toss my black-rimmed sunglasses on to the dashboard and reach for my radio to confirm my arrival. I switch my status to 'in attendance', so that I can't be assigned to any other calls.* I hop out of the car,

---

* Other statuses include 'mobile to incident', indicating that you are travelling to an incident, and 'mobile available' to indicate that you are travelling in a vehicle and available for other emergencies.

burning my hand on the metal rim of the door. I wince in the bright sun as I pull on my boots. I can't be the pretentious officer striding on to the scene in sunglasses, so I opt for the more humble hand-over-the-eyes-and-squint approach instead. I head to the command unit, ducking under some cordon tape, as firefighters rush past me, dragging hoses behind them.

Black smoke is billowing from the roof of a depot. I know that it contains nearly £2 million worth of stock for some of the nation's biggest electrical retailers. Flames are tearing through the left-hand side of the building and licking the outside walls. The fire began in one of the distribution lorries that was backed into the building in one of the loading bays. It has now engulfed six of the nine lorries and a substantial part of the building. If they had had sprinklers, the damage would have been minimal.

I am the most senior officer on the fire ground and I am here, in the first instance, as a mentor for the incident commander. However, I also need to be ready to take over if – though in this case I suspect it will be when – the fire grows. I won't step in until there are at least eleven fire engines and, at present, there are only ten. I pass a firefighter who's been inside already. Her helmet is black and she has soot smudged across her face. There are still so few women firefighters – less than 5 per cent in the UK – that I'm always pleased to see women coming up through the ranks. The outdated stereotype of a firefighter is so pervasive that many women don't even consider themselves in the role.

'How is it?' I ask, hoping for some insight into the conditions.

'Fucking hot, boss,' she replies with refreshing honesty.

I smile and pass her a bottle of water from the pocket of my fire leggings.

'Stay hydrated,' I say. 'It's boiling today, whether you're inside or out.'

She nods and gratefully gulps down the water. She wipes sweat from her forehead and tucks her sooty hair behind her ear.

'Thanks, boss,' she murmurs.

I arrive in the command unit and take off my helmet. I too am sweating and I've not been near the fire yet. I'm surrounded by the dry crackle of radios, and walls with whiteboards covered in every colour of marker pen. I feel a gentle hand on my elbow and turn to be greeted by the incident commander. I know Les Brightman well and have worked with him many times before. I like him a lot and I trust his judgement. He has a bright ginger moustache that spreads across his face and curls into points at either side of his mouth. I've often teased him that the wax he uses on it is a fire risk in itself and needs to go up on the hazard board.

He takes my hand, grinning broadly, and shakes it so that ripples run right through to my feet.

'Ma'am, I'm glad it's you and not that other . . .'

'Les.' I clear my throat.

'You're not going to psychoanalyse me again today, are you?' He grins.

'Believe me, Les, it'll take more than a doctorate to get to grips with you. Now, what have we got?'

Les briefs me, explaining that the fire is spreading internally, and because it is jumping from lorry to lorry, it's now making its way across the outside of the building too. Most of the staff are out of the building, but we are still missing two: the driver of the first lorry to catch fire and a second colleague, who we think went in to look for him.

I admire the bravery of those who face fire willingly to save someone they care about, but I really wish they wouldn't. It's not like in the movies. You cannot battle through the burning corridors and emerge as a saviour, a dark silhouette against a backdrop of bright orange. The reality is far less appealing. After one or two lungsful of toxic smoke, you would fall unconscious. The hot, lethal gases would continue to burn your windpipe and poison your body with carbon monoxide and cyanide until you could no longer breathe. You would likely asphyxiate long before the flames reached you.

Les has split the depot into two distinct sectors, each with its own commander reporting directly into him. There are crews inside looking for the missing people and trying to stop the fire spreading to the high-value stock. There are also crews outside, trying to extinguish the lorries and prevent further spread. There is a store room filled with cylinders to power the forklift trucks and Les has instructed crews to remove the cylinders immediately, so that if the fire does spread, the risk of explosion is reduced. Les delivers a clear, structured and comprehensive briefing. I feel relieved; I think we're in the best possible position and that things will go well.

Our biggest problem here – and something to keep an eye on – is the water supply. Les has asked the water company to increase pressure to the mains, but the volume of water coming from our firefighting hydrants is pitifully low. That means we have fewer jets than we'd like.

I take advantage of the current state of calm and walk

around the fire ground to build up my jigsaw. I've only seen the blaze from one side and I really need to see the whole site. I pull on my helmet, fasten the strap, and step into the muggy summer heat. Sector 1 is the main scene of external firefighting operations and covers the front section of the building. The sector commander is talking into his radio and greets me with a nod. Then he approaches, holding out his hand.

'Sorry, ma'am. Just had to get that. I'm John Fielding, the sector commander.'

He is slight with a slim face and dark brown eyes that almost look black under the shadow of his pronounced brow. He has a broad jawline peppered with a five o'clock shadow.

'No need to apologize,' I reply. I introduce myself. 'I don't think we've met before.'

This isn't uncommon at a large incident. There are over seventy firefighters and officers here.

John outlines the situation in Sector 1. He is focusing on external firefighting and trying to extinguish the lorries. However, his efforts are being hampered by the lack of water. He has the maximum number of jets that the water supply will allow. Despite being visibly frustrated, he too seems to be coping well.

As we're speaking, his eyes dart past me and he points at the lorries. A shimmering trail of diesel has seeped from beneath one vehicle and is running uncomfortably close to another. It would provide a perfect pathway for the fire to spread.

'Boss. A fuel leak. From the lorry. Bear with me.'

He radios another officer. 'We need to extinguish the fire in these lorries,' he says. 'And we need to switch to foam. The contamination risk is small and limiting the ignition potential is more of a concern,' he declares.*

John's behaviour demonstrates intuitive decision-making at its best. He hasn't considered a number of options, weighing up the pros and cons of each, but responded instinctively to a visual cue; in this case, the leaking fuel. He has then used the decision controls to validate his intuitive response, clarifying his goal, his expectations and the risks in order to rationalize his gut feeling. An intuitive response doesn't need to be a blind kneejerk reaction, as John has demonstrated. I'm impressed by his decisive approach. He didn't look to me for approval; he knew he was within his own limits of authority. There was no inaction.

In Strathclyde in 2008, a woman – her name was Alison Hume – fell down a disused mineshaft. She was there for six hours before being rescued. She was profoundly hypothermic and went into cardiac arrest while being lifted to the surface. She very sadly died. The commanders who attended the incident were criticized in the inquiry for inaction resulting in a delay that compounded her symptoms. They did this because they were sticking rigidly to a policy that said they needed to wait for mine-rescue to arrive. The inquiry said that the team was over-reliant on other agencies. Rather than considering other options, there was inaction. Decision inertia through choice deferral, perhaps.

---

* Firefighting foam is toxic to aquatic plants and animals. When we use it, we have to be very careful that it doesn't enter drains and watercourses. However, it is particularly effective for extinguishing fires – especially oil-based fires – and so we must always balance the environmental risk with other risks at the incident.

Neither inaction nor a lack of decision-making seems to be a problem at the incident I'm attending, so I move on. I speak to the crews at the front of the building and feel satisfied that their understanding of the situation matches John's, which in turn matches Les's. I make my way around the perimeter of the sector, getting a sense of the dimensions of the incident ground, so that when I return to the command unit I can quickly and easily understand the marker-pen outline drawn on the whiteboards.

The lorries are nearly extinguished, but the fire is still spreading through the building. Thick, black smoke pulses from cracks in the walls and the flames have breached the roof. I'm concerned about the missing people. I turn up my radio but they haven't yet been found.

I reach the second sector which – somewhat unintuitively – is called Sector 3.* The sector commander is giving an intense and animated briefing to a crew commander. Firefighters are scurrying around, fetching more hose, prepping a casualty care area in anticipation of a live rescue, and setting equipment into another hydrant to reach more water. The pitchy screaming of the pumps drowns out all other sounds.

The rear entrance is on the right-hand side of the building, opposite the burning lorries. I step over the three red

* This is because a building is split into sectors in a clockwise direction. When there are only two sectors in operation, the front becomes Sector 1 and the rear becomes Sector 3. If the approach changes and additional sectors are introduced, then Sectors 2 and 4 can be easily added at the sides without disrupting the numbering system. It seems odd on the face of it but it makes sense!

hose-lines – fully charged with water – leading in through the door. Each represents a crew of at least two firefighters. I glance at the entry control point where a plastic board is standing on a metal tripod. The entry control officer is adding critical information to the board, logging the times at which the crews entered, how much air they had at that point and how long until it runs out. He does it quickly and deftly, simultaneously taking radio messages from the crews inside.

There is an emergency team ready with breathing apparatus in case one of their colleagues begins to struggle. The emergency team is always the same size as the largest crew in the building, so that they can match any crew that might need rescuing. There are three firefighters in this emergency team, so I know that at least one of the crews inside is a team of three. So there is a minimum of seven firefighters inside, but possibly as many as nine.

I approach the sector commander, who has his back to me. The crew commander is now talking to a firefighter operating a pump. I don't recognize either of them. The crew commander's arms are crossed defensively and his brow is furrowed. He looks thoroughly pissed off. I can't hear much of their conversation over the noise of the pump, but I make out something like, 'If he ever speaks to me like that again, I swear I'll . . .'

It doesn't sound good. Then the sector commander turns around and I understand the problem immediately. Saul Mortimer is in charge of Sector 3. There are very few people in this world who I really, genuinely, cannot abide. He is one of them.

He is a renowned bully rumoured to have strong sponsors – senior officers with influence – who people say cover for his bad behaviour. Many believe that if they challenge him, their life will get harder or their career will be limited in some way. This is *highly* unlikely to be true, but you're not dealing with the tree, you're dealing with its shadow. This means very few have the face to take him on.

Except me. I once did. Unsuccessfully. We crossed swords over the way he treated a firefighter, but with very little evidence the case against him didn't stick. I've had a number of other run-ins with him, although like most bullies he doesn't have the mettle to bully upwards. Instead he reserves his putrid attitude for those he outranks, abusing his authority.

I would relish the opportunity to take him on again – in the early days of my career I was regularly on the receiving end of attitudes like his.

I sigh and walk over, bracing myself for the tirade that will inevitably follow.

'Saul,' I say, smiling. 'How are things?'

Deep acne scars mark his face like a lunar landscape. Small and stout, he reminds me of a Russian doll. He grimaces and holds my stare, expecting – or perhaps hoping – for a confrontation. However, now is not the time. Not when a business is burning down and people are missing. So I continue to smile and offer my hand, knowing it will wrongfoot him.

He is visibly startled and contorts his grimace into a feeble smile. He stutters as he begins briefing me and he struggles to hold my gaze, always shuffling and looking between his hands and the scene. He speaks formally, loudly – he's clearly very uncomfortable. I can't help but feel a little amused. I resist the temptation to play with him further, though. Again, this is not the time.

I run through my usual questions, confirming his understanding of the situation and checking his jigsaw and his situational awareness. He interrupts the briefing several times to bark orders into his radio.

'Where's that crew? Listen! I told you I wanted another crew!' I watch his face contort as he bares his teeth. His language isn't just abrupt – which is quite normal over the radio – but antagonistic.

Communication is an important skill, but it does require a lot of brain power. This is even more the case for *effective* communication. Excessive amounts of stress reduce your ability to perform, so when you're under pressure, your communication worsens and you're less able to convey your meaning clearly. This means that your listener – in this case, a firefighter about to risk their life in a dangerous, burning building – will be less able to understand what it is you want of them.

Our brains process communication in a very complex way involving a number of different brain regions. Listening requires the ability to process sounds (the auditory cortex) and attribute meaning to those sounds (the frontal cortex and parietal cortex, among others). Producing language involves formulating what it is you want to say (the frontal cortex, temporal and parietal lobes) and then speaking the words (the motor cortex, in your frontal lobe).

Since 2008, Uri Hasson, professor of psychology at the Princeton Neuroscience Institute, has been running experiments exploring the brain on communication – how neural patterns associated with stories, memories and ideas get transmitted to another person's brain. He and his team discovered that when speakers and listeners are communicating effectively, they experience something called 'neural entrainment', in which soundwaves trigger similar brain activity in both the speaker and the listener.

Problems with communication were identified at both the fatal incidents previously mentioned in Chapter 3: the Marlie Farm fire and explosion in East Sussex, and the Shirley Towers high-rise fire. At Marlie Farm, there was no clearly agreed procedure between emergency services for the emergency evacuation. Although the emergency evacuation signal was given, people remained within the area and the subsequent explosion proved deadly. At Shirley Towers, there was a breakdown in communication and officers were, at times, unclear which commander was actually in charge. While these were by no means the defining issues at either of these incidents, they were part of a complex alignment of factors – the sliding doors that had lethal consequences. We have learned from incidents such as these and we do things differently as a result. We appreciate – always, and with absolute clarity – the vital importance of good communication.

Saul is failing to communicate effectively. Not only is his language and tone aggressive over the radio, but he isn't listening to me at all. He talks over me every time I speak, failing even to hear, let alone acknowledge my contributions. It is a frustrating combination of mansplaining and verbal domination, and he is doing his utmost to prevent me from getting a word in.* He lets me speak for, at most, a couple of seconds before turning and barking at

* The term 'mansplaining' is used typically when someone (often a man) describes something (usually to a woman) in a particularly patronizing way. It really, really irks me.

someone else or straining over my shoulder to see something behind me. I'm beginning to feel agitated. I check myself for a moment, ensuring that I'm not suffering from confirmation bias, processing information that only confirms my perspective of Saul as an idiot. After a brief reality check I am certain that my suspicions are endorsed by his behaviours, and there's not a counter-behaviour that I can spot . . . No matter how hard I look for one.

'Saul. Focus. How much of the ground—'

'Yes, we're making progress. When I first got here I—'

I cut him off. He wants to tell me a story – of everything he's done since he first arrived – but I really need to know what's happening now. I try to redirect the conversation.

'Saul. Show me on the board *how much* of the ground you have searched.'

Fed up, I lead him towards the board. If he is failing to answer a simple and straightforward question, then I am very concerned about his ability to convey risk-critical information to those under his command.

This is an instance in which someone's behaviour – their disposition, their language and their actions – reveals something of their thoughts. I have witnessed poor communication, aggression, an inability to read and respond to the emotions of others, an inability to focus and, to be frank, an inability even to string a sentence together.

I opt for a tactical withdrawal to the command unit. The fire is not yet big enough to warrant my command, but there is a command issue that needs addressing. I need to speak to Les. Saul needs to be supported or replaced – and quickly.

I climb up the steps and back into the command unit

where I am immediately struck by the shift in mood. It had been a calm, serene space but it is now chaotic and full of people, all vying for Les's attention. The radios are loud, drowning out the voices. And the voices are getting louder, trying to drown out the radios.

Les is trying to have two conversations at once and, understandably, he's looking incredibly flustered. He has no thinking space whatsoever. He hasn't recognized the pressure building up. We've all been in that position at some point. It can happen incredibly quickly and before you realize it.

I push through the crowded unit – it's like a tube station in rush hour – to reach the command support operatives. I instruct them to turn down the radios and clear the unit of all non-essential personnel. I need them to then go quickly through everyone waiting outside to speak to Les, find out what they want and what information they have, and prompt us immediately if it's critical to the safety of firefighters or our operational tactics. This will create some thinking space for Les and will ensure that the most important information is prioritized.

The command unit is cleared within two or three minutes.

Les sighs. 'Where did everyone go?' he asks, looking perplexed.

'It looked like you could do with a breather,' I reply.

'I was swamped.'

A command support operative pokes his head back inside. 'There's a few people still waiting. Are you ready for them?'

'Anything risk-critical?' I ask.

'Nothing immediate. Only several officers who've just arrived and need a brief.'

'Thanks. Will you tell them to hang on, please? Put the external board up outside and give them an overview while they wait. I need five minutes first.'

'Ma'am,' he replies and ducks outside again.

I turn back to Les. 'There are some issues in Sector Three, but first, where are we at? What's changed?'

'Uh . . . Well, we've got more resources now . . .'

'Great. How many of the ten pumps are here?'

Les looks blank.

'All of them, ma'am,' says the command support operative.

'Great. And officers?' I ask, looking directly at Les.

He begins to count on his fingers.

'All of them, ma'am,' says the second command support operative. 'Three are outside waiting to be briefed.'

'Thanks. And what are you planning to do with them, Les?'

Les inhales. He looks away, up and to the right, his eyes darting from side to side. He hasn't planned this yet and he's building a possible command structure in his mind.

'Talk me through your plan, Les,' I say gently, hoping to encourage him.

Cognitive planning is a really important skill, particularly in high-pressure, emotionally charged situations. I need him to consider his goal, choose the appropriate actions, and then work out where to begin.

I need to know that Les is still focused on the overall goals – finding the two missing people and limiting the spread of the fire – and that he has a plan in place to achieve these aims. He needs to know who is going to do what and when it should be done. He should have a working jigsaw in his mind, and he should be able to move around it – imagining the jigsaw in five minutes' time, ten minutes' time, half an hour.

However, his behaviours aren't filling me with confidence. I think he knows what he *wants* to achieve, but isn't able to work out a route through his jigsaw towards that end goal.

'OK, Les,' I say. 'Things have got very busy very quickly. We'll both feel better if we recap the situation.' I don't want to blow out his candle.

I walk over to the boards and point to the list of objectives.

'Talk me through them, Les. What were they before and what are they now? What's changed?'

'The priority is still life,' he says. 'We've confirmed that we have only one person missing. The driver. The other guy had left the site.'

'Great. So, only one person missing. That's progress.'

I ask the remaining command operative to update our board as well as the external one.

'What are we doing about the missing person?' I ask.

'That's being managed by Sector Three.'

'And what progress have they made?'

Les pauses. 'They're still searching.'

'Do we know how much of the building has been searched? And how much is left to search? How quickly they're going through breathing-apparatus wearers? Have we got enough crews or do we need more down there?'

'I'm not sure,' Les replies.

I turn to the command support operative. 'Did the last update from Sector Three have an update on progress or resourcing requirements?'

'No, ma'am.'

'Did you ask for it?'

'Yes, ma'am.'

'And?'

'Well . . . To be honest . . .' He looks cagey, as if he has something to say but doesn't feel comfortable saying it. His response confirms my suspicions. Saul's failure to communicate is impinging on our situational awareness and it's definitely not confirmation bias on my part.

'Can you try again, please?' I ask. The radio operator reaches for the radio.

I turn back to Les. 'I'm concerned that Saul isn't coping. I couldn't get a sense of how much he had searched either. I think he needs to be supported or replaced.'

Les nods.

'I suggest that one of the new officers becomes the search coordinator. They can work within Sector Three. They can focus on finding the missing person and reduce the pressure on Saul.'

Les nods again.

We continue to work down the list of objectives, figuring out what has changed and what the next steps should be. We then move on to the list of the risks, and thankfully nothing new has come to light. It only takes seven minutes for Les to feel back in control.

I stand back and Les opens the door, ready to face the madness again.

Les's behaviour clearly indicated that his brain was struggling to process the onslaught of information and respond effectively. These changes in his attitude, body language and demeanour prompted me, as the most senior officer on scene and his mentor, to step in. I assisted him by creating some thinking space – literally and figuratively – so that he could calm down and regain his composure. It is important that managers and leaders can read their teams. In a high-risk situation, the ability to do so can be the difference between saving a life or – frankly – being complicit in a fatality.

Our research identified how important this skill was for commanders, but in order to really make a difference we needed a way for leaders to be consistently aware of and engaged with their teams' behaviours. It needs to be systematic. It's no good preventing someone from heading for a fall once in a while; we needed to catch everything. It also needed to be simple – something that everyone could do with a little training and didn't require a psychologist's skill and experience.

Phil Butler – my PhD student and friend at Cardiff University – Rob Honey and I started working on a method that would do just that. Phil and I had worked together regularly, having originally been part of the team tasked with writing our national policy for incident command. We knew that other fields – aviation, surgery, nuclear energy, oil drilling – had used behavioural marker schemes to reduce human error, so why shouldn't we? Over the course of around six months, we interviewed nearly eighty commanders, at all levels, from across the country, asking questions about how they operate in a crisis: the way they think, the way they feel and the way they behave.

When we completed our interviews, we began to analyse the data. We looked for recurring patterns of behaviour amongst officers and we examined the ways that commanders responded to those behaviours. We then put together an expert panel of experienced commanders to help distil the broad list of behaviours and responses into something more specific. We wanted to identify the behaviours demonstrated when an officer or commander is responding *effectively* and then create an early warning system for human error.

This would become the bones of the behavioural marker scheme. We would eventually have a list of clearly defined behaviours associated with essential skills. The six main skills were leadership, decision-making, communication, personal resilience (operating under pressure), situational awareness and teamwork, and there were also twenty sub-skills.

Then the real work started. We needed to be sure that the list was valid, that it really did identify the right skills and the corresponding behaviours. We also needed to be sure that the list was reliable. We needed to know that different people could use it to determine someone's effectiveness and come to similar conclusions. It also had to be sensitively worded. It wasn't about 'good' or 'bad', but about grading a commander's or officer's performance without bias.

We asked our commanders to watch dozens of videos of

Main skills	Sub-skills
Assertive, effective and safe leadership	Setting and maintaining standards of performance
	Values and supporting others
	Leadership style
	Competence
	Safety leadership
Effective decision-making and planning	Intuitive decision-making
	Analytical decision-making
	Planning
Interpersonal communication	Listening
	Communication style
	Briefing
Personal resilience	Thinking time
	Stress and fatigue management
	Confidence
Situational awareness	Information gathering
	Understanding information
	Anticipating incident developments
Teamwork and interoperability	Cooperation
	Team formation
	People-oriented

officers and commanders at incidents and to rate their command skills using this system. We compared their marks, amending the lists of skills and behaviours and repeating this exercise until we were confident that we had a method that met our criteria.

Phil named the list THINCs (**TH**e **IN**cident **C**ommand **S**kills Behavioural Markers). It has since been adopted by the National Fire Chiefs Council, and officers across the UK are being trained to use a simple app that can record behaviours and systematically score command performance. It will help us raise awareness of behaviours to look out for, and will help us train commanders to better function in pressurized and stressful situations.

I'm back in Sector 3. I took over as the incident commander seven hours ago, when Les increased the number of fire engines from ten to fifteen. It's just gone one o'clock in the morning. It was too long to be in charge, really, but I was emotionally invested. Thanks to the construction of the building – in which a brick partition wall acted as a fire break – we managed to stop the spread of the fire. The front part of the building and most of the lorries are completely destroyed, but the rear is still standing and much of the stock is untouched.

A black body bag lies on the concrete floor in front of me. The stiff sheeting creates sharp ridges that hide the shape of the corpse beneath. The lorry driver didn't make it out alive.

I shake my head. A familiar gentle hand touches my elbow. Les. He has been my operations commander for the last few hours and has done a brilliant job.

'We can't save 'em all,' he says, patting my arm.

'I know,' I say. 'But every one of them hurts a bit. There's someone at home now, expecting him back any minute. Maybe waiting up. Kids in bed. They're about to receive the worst news of their life. After a normal morning waking up to cornflakes and their usual, mundane routine. Bet they'd give anything for mundane routine right about now.'

Les nods. He's been where I'm standing. He knows. 'Poor fuckers,' he says.

Poor fuckers, indeed.

I arrive home at six o'clock in the morning. I open the front door as quietly as I can, shushing the little bald dog, Jimmy Chew, who's known to bark ecstatically whenever someone enters. I bring with me the smoky smell of fire. I usually go straight for a shower, but instead I creep into Gabriella's room. She stirs as I kiss her on the forehead.

'Mum, you stink,' she murmurs, and rolls away, pulling the blanket over her shoulder. I smile.

Somewhere else, in another part of the city, there is a woman, newly widowed, weeping at the doorway of her children's room.

Our every day is someone's worst day. I will never forget that.

10

Hindsight is Twenty-twenty

'Hindsight is good, foresight is better.'

Evan Esar

THE LIGHT above my desk is harsh and fluorescent. I have been reading for hours and my eyes sting. I blink several times. A large, single-paned window spans one wall. Its frame is speckled with dirty white paint flakes that are peeling away. I look outside. Sleet is falling heavily on the cars, coating the yard in frozen sludge. I am relieved to have a radiator beside my desk, pumping out hot air and warming my legs.

I take a sip of hot black coffee. Then I pick up my pen and flick through the stack of pages in front of me: print-outs from the message log, collated debrief forms from the crews, written risk assessments, contemporaneous notes from commanders and an incident summary from the assurance officers. There was a large fire in a printing works – it destroyed the entire building – and I will be chairing the debrief due to take place shortly.

Our daily business is high risk and no two fires are ever the same. It's important that we take the opportunity to learn from each and every incident and from each and every decision. This debrief will be attended by a number of commanders. Each, at some point throughout the incident, was responsible for the entire scene. I need to work out the order of events. I want to understand the reasons behind the actions they took, individually and collectively. Did we do everything we possibly could to save that building?

I'm using the information in the paperwork to build a picture of the incident. I'm looking for two subtly distinct things: evidence and perspective. The hard evidence – including the times on the message log and the written risk assessments – provides a scaffolding. This framework will be expanded using detail from each person's perspective, such as their notes and their recollections in the debrief. I will shortly have the opportunity to ask any number of open-ended questions. These will – I hope – encourage the commanders to offer new information on what they understand happened and their involvement in the response. I will need to find a path through the contradicting memories, the conflicting emotions and the narratives that have been reconstructed in retrospect.

I'm looking for the truth. And it can often be difficult to find. Commanders rarely misrepresent the situation intentionally, but memory is notoriously fallible and people are complicated.* The hard evidence reveals that the fire grew rapidly. Three fire engines responded to an automatic fire alarm in a commercial printing works and arrived very quickly, the first just six minutes after the call was received and the third just three minutes later. The first fire engine confirmed that there was indeed a fire in the building and immediately requested additional help from a further five pumps. The three fire engines sent as standard to deal with the fire alarm weren't going to cut it. Within ten minutes, the number of pumps required had increased to ten. This is concrete evidence that in the first sixteen minutes after the call was received there were two critical decisions relating to the size and ferocity of the fire and to the resources required. It's not uncommon for a serious fire to escalate quickly.

Was that first commander examining the fire as it was

---

* Live footage would allow us to revisit the scene as it really stood and observe the response as it unravelled. However, while we have this option occasionally, we rely primarily on a combination of records, memory and perspective.

then? Or were they thinking ahead, and making a decision and requesting additional resources based on what they predicted that fire would look like in thirty minutes' time when the resources would actually arrive?

Our research shows that commanders typically make decisions based on what is happening in the 'right now' rather than thinking of the 'what will be'. This is not at all surprising given how the brain responds under pressure. However, it's not the most effective way of resolving an incident. This is something the decision controls targeted, and those who use them are more likely to focus on anticipating what might come next.

I am fairly confident that at the sixteen-minute point only three fire engines had actually arrived at a scene severe enough to warrant ten. I think that the initial request for additional support was in relation to the 'right now' and not the 'what will be'.

I need more detail, but there is little in the post-incident notes that covers this precise period. I suspect that things were unfolding so quickly that the information was difficult to process. I pick up my phone and click through to Twitter, looking for photos that were shared in the first half an hour of the fire. There are plenty from later that evening, when the building was completely ablaze, but only one from the early stages. It was tweeted an hour into the incident, but I think it might have been taken earlier. Smoke is billowing from the top-floor windows and flames are breaking out of the black-tiled roof. I download the image and email it to myself. I switch to my desktop and open it up, but the properties tab – where the date, time and location are sometimes stored – is empty.

I flick through the message log to find the first informative message.* It says that at seven minutes and thirty-five
seconds into the incident there was a confirmed fire,
involving the third floor and approximately 20 per cent of
the roof. It isn't a perfect informative message – more
details would have been beneficial – but time would have
been limited and it at least provided the incoming crews
with a first impression of the situation and an understanding of what to expect.

The photograph corroborates the information in the
log. I enlarge it on the monitor for closer inspection. The
printing works was based in a Victorian brick building
that spanned three floors. An old, hand-painted ghost sign
preserved on the side advertised tobacco. My photograph
shows the north-west aspect of the building as the sun was
beginning to set. It was probably taken around 6 p.m. I
check the weather on the date of the fire: heavy rain
throughout the incident, which means damp, cold firefighters. I make a note to check the welfare arrangements.

The log reveals that a canteen unit was ordered – a cup
of hot tea at a long, dirty fire does wonders for morale – but
the hard evidence doesn't offer any detail. I need to know
where the entry control point was established, and where
firefighters were held prior to being committed. It should
have been somewhere dry, where firefighters could shelter
before going into the building. If their kit gets wet before

---

* Informative messages are logged throughout an incident and are
designed to provide a snapshot of the firefighting activity that is taking place.

they enter, the rainwater then turns to steam in the heat and can burn their skin underneath.

I squint at the photo, trying to decipher what exists in the dark shadows. I can't see inside the windows. A few people – onlookers – are standing near to the building. There isn't yet a cordon. The initial crews would have been overwhelmed and trying frantically to respond appropriately with less than a quarter of the necessary resources. How would I have felt had I been the initial commander? What would I have done?

Jess Owens, the first commander on scene, had been called to an automatic fire alarm. They usually turn out to be nothing. She arrived, expecting a non-event, but was confronted by a serious fire. There would have been an immediate conflict of expectations. She would have felt surprised; she wasn't necessarily prepared for a busy, frantic scene and such a quick response. Her brain would have been dealing with competing priorities and countless pieces of disparate information, and trying desperately to integrate everything into a coherent picture. At the same time she would have been attempting to put in place the textbook measures for such a situation. The difference between the reality and the theory is crucial. I need to understand what Jess was thinking in those first few minutes before I can judge if her actions were good, bad or indifferent.

The role of the initial commander is critical. They set the tone for the incident and their decisions are the foundations upon which every other decision is made. A misjudged action or a missed piece of information can have serious ramifications further down the line. The actions of an initial commander can be the difference between serious escalation and an early stop.* Even a fire engine parked in the wrong location can cause a serious headache later on if the incident escalates and it is stuck in position providing water to a specific crew. However, sometimes the outcome has already been set by circumstances beyond any firefighter's control: the fire might be too severe or the right resources might not be available.

I trawl the message log. I need to know who took over after Jess. When did they arrive at the incident? What time did they take command? What else happened before they took command? How long did they stay in command? All of this information is essential: it tells me if a commander was already on scene developing their situational awareness before taking responsibility, if their jigsaw had the potential to be excellent (or only ever mediocre), if they had the opportunity to make a difference, and – sometimes – if they were fatigued or exhausted after a particularly long stint.

Jess, a crew commander, was succeeded by Stan, a watch commander, after just six minutes. Jess therefore had very little time to develop her situational awareness before she handed responsibility for the incident to Stan. She would barely have had time to get started.

Stan had a minute or two to familiarize himself with the scene before taking command, and then was in charge for a grand total of twenty-two minutes. He was replaced by a

---

* A 'stop' refers to the 'stop message' which signals that we don't need anything further at the incident. It is only sent once the fire is practically out and we are damping down any final hotspots and closing the scene.

station commander, Mark Potts, who held the reins for the next sixty minutes. Eventually, a group commander, Tony Lewis, arrived on scene at 19.28 and took over responsibility for the incident.

When Tony arrived, the fire was still raging, orange and red flames flickering violently against the sky. There were ten fire engines and their crews were valiantly battling the fire to stop it in its tracks. The foundations of the response had already been decided. Tony managed the incident for the final eight hours. In total, there were four incident commanders over a period of almost ten hours.

It is also important to examine the resources that arrived – both fire appliances and officers – throughout each commander's time in charge. Did they have the necessary equipment to action their piece of the plan? Had their predecessor thought ahead? When Stan first arrived, he had only three fire engines. However, when he handed over responsibility to Mark, there were eight trucks in total, with a further two on the way, and all had been briefed and were already hard at work. A further five pumps – with crews and three other officers – arrived throughout Mark's command and he briefed them and allocated their responsibilities. Tony reviewed the activity of all allocated crews, and received a further five officers. How did each commander brief their crews? How did they use their allocated resources? Did they make the right decisions?

The crews who served under each commander have completed debrief forms. From each commander I will be getting a first-hand account at the debrief. I need an understanding of the crews' perspectives in order to analyse whether all the jigsaws matched. So often they don't. That's why 'ground truth' – what you can directly see to be true – is so important.

If, as a commander, you stay on a command unit – sat away from the incident and with only descriptions rather than real visuals – for too long, then you lose sight of the reality. You are imagining an incident that may or may not be accurate. You aren't experiencing the harder, rougher version that your teams are encountering.

Periodically looking up and out of the metaphorical tunnel, physically getting off the command unit and putting your boots on the cold, hard ground is so important. It is often hard to find a lull in activity, a natural gap in the response, so sometimes you simply have to make one. It's OK to be creative. Walk and brief simultaneously if you must, but for the sake of those trusting you with their lives, get up, get out and get with it.

The debrief forms reveal plenty of conflicting details. The crew members disagree on what was briefed, where and when, with different locations, times and sequences of events in different reports. I need to tease out which are errors of memory – people forgetting the precise details – and which are real errors, mistakes that occurred in the moment (the result of a poor decision or a failure to process information) that may have reduced situational awareness.

I glance up at the clock. I reckon I have just enough time to make another coffee before the teams arrive for the debrief. I always prefer to review the material immediately beforehand so that the information is fresh in my mind.

There's a knock. Brian, one of my officers, pokes his head around the door. He's as wide as he is tall, a mountain of a man, ex-military, with short, cropped dark hair.

He might look intimidating, but he is a gentle giant, razor-sharp and very efficient. I like working with him.

'Ma'am, they've all arrived, albeit a little early. Are you ready for them now or shall I tell them to wait?' he asks.

'It's OK,' I say. 'Early is fine.'

He glances at my empty mug. 'More coffee?' He grins. He knows me well; it's a rhetorical question. 'I'll bring a fresh pot.'

I gather my papers, zip them into my worn leather folio case and make my way down to the main meeting room.

I open the door and the chatter subsides immediately. I smile and greet the room. Many opt to chair these debriefs in a very formal manner, but I find that the conversation is more productive when participants feel at ease. There are seven people waiting to take part in the discussion. I haven't met most of them before, and they all look apprehensive, which isn't uncommon. It's natural to anticipate criticism – though not always needlessly. Sometimes criticism is warranted, but all too often it can be personalized. I firmly believe that you'll never encourage an open and honest conversation if someone is afraid of being embarrassed or harshly criticized. It stifles psychological safety and makes people feel disrespected and alienated. It prevents them from opening up and sharing what might feel like a vulnerability, which if it remains masked can never be addressed.

I notice that no one has a drink. Embarrassed by the idea that I might rudely be the only one with a coffee on the way, I ask Brian to call for some refreshments. I tell the commanders to talk amongst themselves for a few minutes while we wait. This isn't just a random act of kindness, however. I can use the brief pause to see how everyone is interacting. The people in the room are, in effect, the command team who dealt with the fire in the printing works. I don't yet have a sense of how established their relationships are, and watching them now might help me to understand how they interacted on the night. I want to know who knows who and where frictions might exist. I sit quietly and observe.

The room remains silent. I turn to Brian. I open my mouth and he grins again. He knows exactly what I'm doing.

'Good weekend?' I ask.

'Average,' he replies. And the murmur of voices begins.

I pull out my documents and arrange my notes in front of me, alongside a blank pad of paper and a pen. The relationships are becoming apparent. There are two group commanders: Tony Lewis, the final incident commander, and Joshua Wattle, who was the on-scene operational assurance officer. Joshua's role was to observe the quality of the response – was everything running as it should or did practices slip? – and to be the commander's eyes and ears across the incident site.

Tony and Joshua are making small talk, but I sense unease. Joshua seems slightly deferential. Mark, the station commander, attempts to break into the conversation, but neither one of the two senior officers is particularly receptive. Stan and Jess, the first two commanders, are deep in conversation. They know each other well and laugh together like old friends.

I can sense a hierarchical dimension to the relationships within the group and I wonder how power dynamics may have affected the response. Could the hierarchy have influenced communications between different ranks? Did everyone feel empowered to share their suggestions and situational awareness? Or were more junior team members silenced, pushed down into their place? I have nothing definitive on which to base my hunch, and I don't intend to introduce a new confirmation bias, but I make a mental note to visit this misgiving as a line of enquiry.

The drinks soon arrive and everyone pours themselves a cup. I can begin. I call for the room's attention and we go around the table, introducing ourselves and establishing everyone's role at the incident. I plan to take each person through their account, from the moment they were alerted to the situation to the point at which they handed over to the next incident commander. I start with the first two on scene.

It's clear that the first twenty to thirty minutes were very hectic. Jess barely had enough time to confirm that it was an actual fire rather than a false alarm. Stan established the fire's location and instructed the first crews in breathing apparatus to enter the building while securing a water supply from a nearby hydrant. There was no command unit at this stage – nowhere that offered sanctuary from the madness or an opportunity to deliberate quietly – and there was more for Stan to do than would have been physically possible within the time frame of his command. Over the course of the incident Stan had the most to do, the most limited resources and the most pressure. I think that both he and Jess did a good job under the circumstances.

I pose a few further questions to both Jess and Stan before opening up the conversation to the floor, allowing others the opportunity to ask any follow-up questions. Reliving something so emotional can feel cold, clinical even, during a debrief. I'm mindful of how it can be to relive stressful experiences – some of the team might be holding on to professional regrets – and it's important to be sensitive to this. Mark has nothing further to add; neither does Joshua. Tony asks if he considered sectorizing the incident and Stan looks almost offended. Given his small window of command, I'm not sure I blame him. I suspect the question was more about Tony's ego – just to have his voice heard in the room – than the pursuit of the truth. I remind everyone of Stan's limited time frame and we move on.

Mark is incredibly nervous. He has been promoted recently and has never before appeared at a debrief for a fire of this scale. He explains where he was when he received the call. He had just got home from the fire station, given the kids a kiss and popped a pizza in the oven. No sooner had he switched from work mode to home mode than he needed to switch right back again. It was a short journey to the scene and he was confronted by a fire that was taking hold quickly and escalating rapidly.

He articulately, but uneasily, explains what his overall plan had been. He had wanted to stop the fire from getting any worse. He knew that there were no people involved, which was some comfort, and he believed it was possible to prevent the building burning to the ground. He had been working to control the fire, using crews inside the building.

Tony, the group commander, arrived about half an hour later and pulled the crews out of the building, opting instead for a 'defensive' approach and fighting the fire only from the outside.

I sense friction between Mark and Tony.

Tony begins to question Mark's account. He is particularly focused on the 'weight of attack'. This refers to the strength of the response: how heavy duty the tactics are and, in particular, how much water can be pumped on to the fire. For smaller fires we tend to use hose-reels, which are very light and easy to manoeuvre. However, if we need a greater weight of attack, we use larger jets – either 45mm or 70mm in diameter – which carry more water. In the case of the printing works fire, which grew very quickly, 45mm jets would have been the most appropriate weight of attack. Hose-reels would have struggled to deliver enough water to subdue the fire.

'Mark, what did you send the initial crews in with?' Tony asks.

'Two forty-fives, guv,' Mark replies.

Tony raises his eyebrows. He looks at Joshua, who nods and says he also saw hose-reels.

'I saw a hose-reel,' Tony says. 'You sent them in with hose-reels.'

'No, guv,' replies Mark. 'It was definitely forty-fives.'

'I saw hose-reels going inside the building. You sent hose-reels.' Tony is defensive.

We all know that the fire grew quickly. Tony believes this was because there was not enough water available to fight the fire aggressively. He thinks Mark is at fault. His question was loaded. Hence the tension.

Mark looks dismayed. 'No. I can assure you, I only sent in forty-fives.'

'Well,' says Tony. 'I saw hose-reels and so did Joshua. So are you suggesting we're both liars, then?'

'No!' Mark protests.

I intervene. 'OK,' I say. 'That's enough, thank you. I think we've established that Mark believes he used forty-five jets, and you believe they were hose-reels.'

We have a conflict that is directly relevant to the course of action taken and I am responsible for unpicking it. Tony is right that more water would likely have improved conditions and prevented the fire from spreading. Whether it would have been enough to ultimately save the building is too difficult to say.

I turn to Tony. 'When did you move from an offensive to a defensive mode of operations, Tony?'

He pauses, blank for a second, realizing that I've made the connection before he can string out the narrative.

'Pretty much straight away, ma'am,' he says. 'It was clear that we'd lost it and—'

I cut him off. 'So you arrived at 19.28. How long, do you think, before you took charge?'

'It was a very quick handover. I could see the fire was out of control and I was concerned about safety. I'd say within ten minutes of me arriving.'

I turn to Joshua, who was there at the time. 'Would you say that was right?' I ask.

'Yes, ma'am. I'd say that was spot on.' He nods his head and glances over at Tony. I sense a camaraderie developing between the two – aided no doubt by Joshua's deference and Tony's need for support. I suspect it might be skewing their recollections.

'And in your view, Tony, was it escalating so quickly that it warranted such a quick handover?'

'Yes, ma'am,' he says smugly.

'And in what way were you concerned? What could you see?' I ask.

'The fire was spreading. I'd say about seventy per cent of the roof was alight. We weren't coming back from that.

There was no life-risk involved and I thought we were going to lose the building, which we did. I could see a hose-reel going into the building so I knew they hadn't used the correct weight of attack. I thought it was too late and, to be honest, I was worried about other possible mistakes.'

Mark looks crestfallen.

'So you believed there had been an error of judgement? And that led you to believe that a quick handover was imperative? More imperative than getting a full view of the situation, so that you could put these errors right?' I ask.

'Yes, ma'am. That's correct.'

I turn to Mark.

'Is that your recollection too?'

Mark shuffles nervously in his chair. I can tell he wants to say something but doesn't quite feel able.

'If that's what the guv remembers, then it must be true,' he answers quietly.

'I'm interested in what you remember. Tell me what happened when Tony turned up.'

Mark looks sheepishly at his hands, his fingers fidgeting with each other.

'Tony arrived and I gave him a quick briefing. He then went to get a good view of the incident ground and returned maybe half an hour later to take charge.' He continues to explain that the tactical mode didn't change immediately and urgently, but confirms that it was the first thing that Tony did as the new incident commander.

I turn to the pile of papers in front of me.

'Tony,' I say. 'You arrived at 19.28. But Mark's recollection is accurate. You didn't take charge of the incident until 20.12, some forty-five minutes later.'

Tony is now wriggling uncomfortably in his seat. He is realizing that his perception of time was incorrect.

'Tell me about the decision to move to a defensive approach. What prompted it?'

'I saw hose-reel jets,' Tony says. 'That's when I knew.'

'And when was that?'

'I can't remember exactly, but I know it was pretty quickly.'

Mark looks affronted again. I ask Joshua for his view and, once more, he confirms Tony's perspective.

I pull out some photographs that were taken by fire investigation officers later that evening, when the fire was nearly out. One was taken at 00.30. It shows the entrance to the building and the door that crews had been using initially to access the third floor. It doesn't show hose-reels going into the building, but several forty-fives. I place the photograph on the table. I rifle through the pile for a second photograph. It shows the same door, but an hour and a half later, when there are hose-reels in place. I put it down next to the first photograph.

I turn to Mark and point to the first image.

'Is this in line with your recollection?' I ask.

'Yes. There were definitely no hose-reels used when I was in charge.'

I turn to Tony. He shakes his head. 'Nope,' he says. 'They must have come out and gone back in again.'

I go back to my file of hard evidence. I read aloud from a debrief form: 'As we were dealing with the last bits of smouldering material, the sector commander told us to go in with hose-reels instead. They were lighter and much easier to manoeuvre and, given the fire was mainly extinguished, there was no need for such a significant weight of attack.'

Mark looks vindicated. Tony is as white as a sheet.

I pull out the resources log. I turn to Joshua and remind him that he was relieved by a replacement officer at 23.44, before the hose-reel was introduced. He clearly did not see a hose-reel in place. He too is now as white as a sheet.

Stan grins from ear to ear and Jess purses her lips, trying to stifle a giggle. My intention is in no way to humiliate or to embarrass anyone, but if I had followed Tony's confident rhetoric, I would have assumed a failure in Mark's decision-making. It would have been easy to do this – there was an independent witness supporting Tony's description of events – and yet it would have been totally unjust.

That said, I'd be wrong to assume that either Tony or Joshua was deliberately trying to deceive me. Our memories can be extremely inaccurate. There could easily have been a problem with how they recorded the incident. For example, part of the memory process which translates – or encodes – experiences into our brains, ready to be called on as memories, may not have happened quite correctly. We know already that a stressful environment can affect brain processing, making information more difficult to absorb and reducing our capacity to process key information. Tony may simply have been too overwhelmed when he arrived to encode and manage all of the details. That might have been why his recollection differed.

Or he might be having a problem with memory retrieval. Tony did see a hose-reel (although, admittedly, much later than he originally stated). It's possible that he reconstructed his narrative subconsciously in order to rationalize his actions, integrating this piece of information in a way that made sense to him. It might even have been a bit of confirmation bias. He might have felt that Mark hadn't been aggressive enough in the early stages of the response, and when he saw the hose-reel later that evening, it fitted with his suspicions and overwrote his original memory. It's like when a document is retrieved on a computer – it can be edited or even overwritten. It isn't static; it can change.

Even for Joshua, this misinformation might have been a factor in his recollections. In a judicial setting, after being questioned by police, it is not uncommon for eyewitnesses to modify their memories of what they saw at a crime scene. This isn't necessarily malicious, or intentional, but a result of the brain's plasticity.

The cognitive psychologist Elizabeth Loftus dedicated her career to investigating this phenomenon and has published studies that demonstrate how memories can be changed. She found that eyewitnesses can be influenced by descriptions they hear after the event, and that this information can affect their ability to identify suspects from a photograph. New information can affect memory, causing us to think we've seen things that we haven't.

Memories can also be implanted. In one famous study, Loftus spoke to the parents of young children and asked them for descriptions of real-life events (moving home, for example) as well as additional made-up events (such as getting a finger trapped in a mouse trap) that had happened in their children's lives. The researchers then asked the children to imagine one of the two kinds of situations, but didn't tell them if it was a real one or a made-up one. More than half of the children in the study generated memories that built on a made-up event. A few, once the truth was revealed, insisted that the made-up event had really happened.

Loftus and her team also demonstrated, in an elegant study, that adults' memories operate in a similar way. Participants were recruited to a programme evaluating the effectiveness of Disneyland advertisements based on the experience of visiting the park, or so they were told. There were four groups. Group A read a relatively plain advert. Group B read the same advert, but accompanied by a picture of Bugs Bunny (not a Disney character) stood outside the Magic Kingdom. Group C read the plain advert but there was a giant cardboard figure of Bugs Bunny in the room with them. Group D read the advert with the picture of Bugs Bunny *and* had the cardboard cut-out in the room.

Each participant was then asked whether they had met Mr Bunny when they visited the park previously. Incredibly, a third of those who'd read the advertisement with Bugs's picture – Group B – said they could remember shaking his hand (or paw, technically). Whereas in Group A – who'd read only the plain advert – just 8 per cent of the group remembered shaking his hand. The cardboard cut-out didn't make a huge difference to Group C (only around 4 per cent had a false memory), but 40 per cent of the double-whammy group – Group D, who'd seen the photograph *and* the cut-out in the room – reported the false memory.

Those people, with just a little suggestion, had created a whole new memory that was totally fictitious, which shows just how vulnerable our memories actually are. Ever been discussing a family event and your brother swears blind that your gran was there but you're not so sure? Or distinctly remember talking to someone at a party and then are slightly embarrassed when they remind you that they were actually out of the country at the time?

Interestingly, Stan had gone on to be the commander in charge of the sector in question when crews took in the forty-fives. He had briefed them to do so. Throughout Mark's protestations during the debrief, however, Stan had remained silent. He had watched as someone in a position of power – Mark – was derided by someone in a position of even greater power – Tony – and backed up by another – Joshua. It's not unreasonable to assume that Stan felt powerless to speak up, or apprehensive about the consequences. I can make that debrief room a safe environment and control the conversation, but once individuals walk out of there they have an entire career ahead of them. Sometimes an awareness of self-preservation prevails.

In an attempt to overcome the apprehension that some might experience in a debrief setting, Jonathan Crego created an ingenious system, known as the 10,000 volts debrief. It allows a trained facilitator to direct a debrief in which participants comment and respond completely anonymously to questions posed by the facilitator or comments from other participants. It strips out power, relationships and the fear of reprisals (or career-limiting commentary!). It allows the quietest voice in the room to be heard at the same volume as the loudest. I've used it previously and I'm sure that it elicits the commentary that would otherwise remain unheard.

You will remember that in an earlier chapter we strapped cameras to firefighters' helmets and, in the debrief, the footage functioned as a visual cue that prompted their recollections of the incident. Some fire and rescue services now use helmet cameras regularly, and acknowledge that they make debriefs more accurate. However, it will take time before they are an established method of practice for everyday use.

Until then, the challenges of memory exist. This is true if you're a watch commander holding a quick 'hot' debrief at the back of the pump before you leave an incident. Or someone like me, chairing a formal debrief.

Or at a coroner's inquest or a public inquiry. Coroners primarily are concerned with identifying the cause of death, including how and when exactly it happened. It is not uncommon for first responders to be called to give evidence at a coroner's inquest where they provide details on what they saw and what they did. If a firefighter loses their life at an incident, it will almost certainly result in a coroner's inquest. Although coroners are not typically concerned with liability, they have a duty to make a report if they believe that action can be taken to prevent a future death.* They will therefore report to fire and rescue services if they believe our practices have gaps – such as a need to review our training, equipment or procedures – and that we can prevent further deaths occurring in the same circumstances.

The more formal the setting of the incident, and the greater the consequences, the more difficult it can be to establish full and accurate memories. The stakes are high and the stress levels even higher. However, in these situations,

---

* Under the Coroners and Justice Act 2009, coroners send out a Regulation 28 report, previously known as Rule 43 reports.

accuracy and truth is imperative. The case is even more so when family members are suffering. They deserve, at the very least, the closure of understanding their loved ones' last moments. Not in pursuit of a case, or a lesson, or a person to blame, but in pursuit of the truth.

So, for all of the forensic analysis of incidents that went right or wrong, our conscience determines the lessons we learn: that foreboding sense of 'what if . . .' and 'if only I . . .' The playback of every permutation of each situation. These things encourage us to develop, to improve, and to be better next time.

With the benefit of hindsight it's so easy to judge a past decision. It is right to judge too, as that is how we might find paths that lead to greater success – more lives saved and better firefighter safety. As firefighters we always judge ourselves. We are first on the scene. We are first to question our response. We will always be first to carve out a pathway for progress. We do this job because we want to help, because we really do care.

One more thing. I have the great privilege of working with some truly remarkable firefighters. But they are only human. We are all only human.

Afterword

It was important to me to write this book. Over the last decade of researching and nearly two decades of fighting fires, I have learned so much and I am so grateful to have had this opportunity to share it with you. The nature of firefighting is that lessons tend to be learned in incidents that are painful, tragic and often life-changing. It is vital that those lessons aren't wasted.

The most significant lesson, for me, has been the importance of empathy. It is the place where I started this book, and it's where I will finish. I have worked to discover the many qualities that a successful firefighter needs, but I truly believe that empathy is the most crucial of them all. It doesn't matter your shape or size, brain or brawn, gender, race or religion – it is empathy that will push you that little bit further than you ever thought you would go. It has been critical for me at incidents, but it has also pushed me to keep asking questions, to put in the hours, to do the research and to balance (or at least try to!) my work with my family life.

Empathy enables you to treat others as you would wish to be treated yourself. A degree of empathy comes naturally to most of us, but to climb inside someone else's mind is hard work. It is challenging to always and consistently be compassionate rather than indifferent, regardless of your mind-set. It is sometimes a struggle to step into someone else's boots. However, it is important to always remember to take off those boots at the end of the day.

Sometimes we give too much of ourselves away and, when we do that, we discover our limitations. Life can be stressful, overwhelming, and often exhausting, and we mustn't take our mental well-being for granted. I think this is particularly true for front-line workers but, for all of us, it is critical that we recognize when we need a little help. And we all, at some point or another, will need a little help.

/ Whether you are wearing a firefighter's uniform, or sitting at your desk in a suit, or sat on a beanbag in your distressed jeans, you are susceptible to the decision traps described in these pages. You are vulnerable to these limitations too. However, now that you understand them, you are more likely to spot them and counter them before they result in human error. You know now how your brain processes information under the extreme pressure of life-and-death decisions. You know how to be more effective.

I'm so proud that the techniques we developed through our research have contributed to making firefighting, and keeping firefighters, safer. They also help us to make sure that members of the public and their property are better protected. However, I know that they only work if we practise them all of the time, which we currently do, and recognize when they aren't working and we need to do more.

We started by looking at human error, and I worked through decision traps and decision controls, through to skills and behaviours. All because we were determined to change and believed in constant and active critique. I hope you might feel just a little inspired to challenge the status quo of what you think you know, peel back the layers of assumption and find new perspectives, new angles and new ideas.

There is one final thing that I would like to add. The reality is that – even with the best will in the world, even with every technique at our disposal – there will always be human error. When that happens, it is impossible to unpick the truth accurately with retrospect alone. When you sit at home and turn on the news or read the headlines in the morning, the right course of action is so obvious. I know; I do this too. I think, 'But surely . . .' and 'How didn't they see . . .' and 'What were they thinking when they chose that strategy?' However, in the heat of the moment, in the thick soup of noise and light and so much information, with the knowledge that lives are on the line, it is not so easy to find that razor-sharp clarity.

Judge us, and hold us to account, and demand an effective, efficient service, but walk a mile in our shoes too. Know that not a day goes by when we don't ask ourselves if we could have done better. Even when, sometimes, our best would never have been enough.

One final, final thing. We have come such a long way, but there is still a long way to go. We are now researching how strategic commanders make decisions that affect entire cities and how we can ensure the best possible chance of the best possible choice. I'm also planning to delve back inside the black-box neuroscience of decision-making in emergencies.

I want to look at biological markers, such as spikes in

heart rate or subtle increases in perspiration, or even eye movements, that might precede a particular type of decision being made, perhaps when we first spot something risky – a sign that might reveal something we've identified before we're conscious of it. I want to understand 'gut instincts' and continue to improve training in the fire and rescue service.

I hope my research will help us to find innovative ways to make up for a decline in on-scene experience, to build the skills that accompany real exposure. Think of it as a gym for the mind and, if we get it right, the results will be applicable to all high-risk industries and sharpen our responses when major incidents hit – and, as recent events in London and Manchester have demonstrated, we do expect them to hit.

I intend to carry on researching and to carry on firefighting . . . and who knows, I may even carry on writing too! Over time, I've learned that you don't need to be defined by other people's expectations. That has given me the freedom I needed to try something different. I love my job. There have been some hugely challenging moments, both professionally and personally, but this is a vocation – it's in my bones. And, even in my very worst moments, it has always been a privilege to be trusted to make a difference on the darkest of days.

Notes

UK Fire and Rescue Service Rank Structure

- Commissioner/Chief Fire Officer
- Deputy Commissioner/Chief Officer
- Assistant Commissioner/Chief Officer
- Deputy Assistant Commissioner/Chief Officer
- Area Commander
- Group Commander
- Station Commander
- Watch Commander
- Crew Commander
- Firefighter

Roles During an Incident

Throughout the book I refer to various roles on an incident ground, all of which are set out below. The list is not exhaustive (to make it so would require a few pages!) but is intended to highlight how we organize ourselves on scene.

Incident commander – In charge of the incident and all the fire and rescue resources in attendance.

Operations commander – At larger or more complex incidents, an operations commander reports to the incident commander and oversees the sector commanders, ensuring that tasks are actioned correctly.

Sector commander – Reports to the operations commander (or directly to the incident commander at small incidents with no operations commander). Oversees the activity in a defined area of the incident ground (a sector).

Logistics officer – A type of sector commander with specific responsibility for overseeing the logistics of an incident, to include equipment, specialist resources, appliances and personnel.

Breathing apparatus entry control officer – Coordinates and monitors the activity of firefighters who are wearing breathing apparatus. They also monitor those firefighters' use of air and calculate when they need to come out.

Command support team leader – Leads the team of command support opperatives in the command unit. Assists the incident commander by keeping all information up to date, logging data and maintaining communications.

Command support operative – Assigned to the command support team. They may undertake a range of activities including mapping the incident ground, logging information, and running radio communications.

Radio operator – Usually (but not exclusively) a command support operative tasked with operating a line of communication with Control (based in a remote control centre) and running radio communications with other commanders on behalf of the incident commander.

Control operator – Works in the control centre that receives emergency calls and dispatches the appropriate fire and rescue service resources to the scene.

Equipment List

Breathing apparatus (BA) – A cylinder of compressed air attached to a face mask, contained within a harness worn on a firefighter's back. Firefighters will wear breathing apparatus when in an atmosphere that is irrespirable (such as smoke, carbon monoxide or toxic gas). Wearing breathing apparatus is hard work, often physically and psychologically demanding, and, in hot and smoky atmospheres in particular, it hinders some other normal sensory perceptions (such as smell – all you get is rubber!).

Pump – A fire engine that has an integrated pump, used to supply water to fight a fire.

Hose-reel – High-pressure hose tubing, usually around 19mm in diameter, used to deliver water (or foam) from a pump to a fire. It is light and manoeuvrable, and is easily pulled off a drum in the rear locker of a fire engine to be used quickly.

Forty-five – A type of firefighting hose that is around 45mm in diameter. It can be used with either high or low pressure to deliver water (or foam) on to a fire. It can deliver more water than a hose-reel, but is heavier and less manoeuvrable. Also available is an even larger hose, 70mm in diameter, which can be used to deliver even greater quantities of water, but is even heavier and incredibly difficult to manoeuvre.

Research Scenarios

In Chapter 7 (page 177), I mention several scenarios that my research team and I used throughout the National Decision

Trials in order to test how effectively our new techniques worked. Here are a few details on these specific scenarios.

House Fire

In this scenario, commanders were faced with a serious house fire. A seven-year-old child is trapped inside by flames. The child's father is at the scene and is absolutely distraught, putting additional moral pressure on the incident commander to act. This scenario is designed to be high stakes (because a life is at risk) and high pressure (because the situation can deteriorate quickly).

We used a number of variations to test how well the commanders would respond. For example, one involved a rapid rise in temperature, indicating a flashover was about to occur, at a point when the commander could hear the child in need of rescue. Injects such as this were designed to present a conflict in goals, so that the commander had to balance their objectives of saving a life and keeping the firefighters safe.

What would you do in this situation? You know the room is about to be overwhelmed by flames. Are you willing to risk the lives of your crew? There might be enough time for someone to dash into that room and pull out the child. You might be able to save everyone. However, there is also a chance that, if you order one of your crews to continue to pursue the original objective, they will die too.

Road Traffic Collision

A car has been involved in a serious collision with another vehicle. A person is trapped inside with serious neck and back injuries. The extrication is particularly difficult because the car is an undercover police vehicle. If you use traditional cutting techniques, you will slice the hidden wires connected to covert surveillance cameras, wiping critical evidence as a result.

This scenario isn't focused on time pressure but on considering potential courses of action. What would you do? Do you prioritize the well-being of the injured person? Or justice? We used a number of variations, such as the unexpected failure of a key piece of cutting equipment, to see how the commanders responded to the unpredictable nature of the role.

Shop Fire

An incident commander is sent to a report of a shop on fire. When they arrive, it transpires that the fire is actually in a skip behind the shop. The site is difficult to access and, bizarrely, they're struggling to extinguish the fire. It seems there is something unusual about the burning material. Very soon, the fire spreads and becomes a serious fire in the shop itself. To add to the pressure, there is an adjoining block of flats. This too becomes affected quickly. What do you do? How do you stay calm and decisive in the face of such unexpected escalation? What are your next steps?

Bibliography

Chapter 1 – Trading Places

'Fire and Rescue Authorities: Health, Safety and Welfare Framework for the Operational Environment' (Department for Communities and Local Government, 2013) outlines the importance of considering human factors in order to keep firefighters safe: www.gov.uk/government/publications/health-safety-and-welfare-framework-for-the-operational-environment

The impact of stress on the way emergency responders process information and make decisions is outlined in more detail in the following book: Flin, R., O'Connor, P., and Crichton, M. (2008). *Safety at the Sharp End: A Guide to Non-technical Skills*. Boca Raton: CRC Press.

There is also a fantastic discussion on how stress affects people working in teams, which can be found in the following publication: Cannon-Bowers, J., and Salas, E. (1998). *Making Decisions Under Stress: Implications for Individual and Team Training*. Washington, D.C.: American Psychological Association.

Professor Rob Honey and I have a small research group at Cardiff University and we focus on factors which affect risk-critical decision-making in emergency responders. We have several PhD students working in related areas. Phil Butler has been focusing on the development of the behavioural marker scheme and Byron Wilkinson has been looking at decision-making in Strategic Coordination Groups. Rob and I have just joined the supervision team for Paula Riella, who is exploring ways to increase people's resilience to PTSD. We are also part of a multi-disciplinary research group, Human Factors Excellence (HuFEx), at Cardiff University. The group comprises internationally renowned academics who specialize in neuroscience, cognitive science, social and environmental psychology, developmental and health psychology. Further information can be found here: www.hufex.co.uk

Chapter 2 – Wicked Problems

The scenario explored in detail in this chapter was developed by a research group led by Professor Laurence Alison at the University of Liverpool Centre for Critical and Major Incident Psychology (with whom Professor Jonathan Crego collaborates) in conjunction with officers from Merseyside Fire and Rescue Service.

More information on decision inertia can be found in the following paper: Alison, L., Power N., van den Heuvel, C., Humann, M., Palasinksi, M., and Crego, J. (2015). 'Decision Inertia: Deciding Between Least Worst Outcomes in Emergency Responses to Disasters.' *Journal of Occupational and Organizational Psychology*, 88(2), 295–321.

Chapter 3 – Only Human

Dr Paul Ekman has conducted a large body of work exploring micro-expressions and emotional leakage. See the 2015 book he co-authored with Wallace V. Friesen, *Unmasking the Face: A Guide to Recognizing Emotions from Facial Expressions* (Los Altos: Malor Books) for more information.

For an overview of confirmation bias, see the following review: Nickerson, R. S. (1998). 'Confirmation Bias: A Ubiquitous Phenomenon in Many Guises.' *Review of General Psychology*, 2(2), 175.

For additional information on mental models, see Gentner, D. and Stevens, A., eds (2014). *Mental Models*. New York: Psychology Press.

More information on tunnel vision can be found in Martin, D. (2001). 'Lessons About Justice from the Laboratory of Wrongful Convictions: Tunnel Vision, the Construction of Guilt and Informer Evidence.' *UMKC L. Rev.*, 70, 847.

The report from the coroner's inquest into the London bombings of 7 July 2005 can be found here:
http://webarchive.nationalarchives.gov.uk/20120216072438/
http://7julyinquests.independent.gov.uk/directions_decs/index.htm

Operational discretion and its applicability to command is outlined in the National Operational Guidance for Incident Command, more detail can be accessed here: www.ukfrs.com/foundation-knowledge/foundation-incident-command?bundle=section&id=17001

Two firefighters tragically died, along with the person they were trying to rescue, at Harrow Court in Stevenage, Hertfordshire in 2005. More detail can be found here: www.hertsdirect.org/infobase/docs/pdfstore/harctreport.pdf

An explosion caused by the mass storage of fireworks at Marlie Farm in East Sussex in 2006, killing two. The following report outlines the significant findings from the incident:

www.esfrs.org/news/2011/july/20110726MarlieFarmSignifica
ntFindingsReport.shtml

Two firefighters lost their lives while fighting a fire at Shirley Towers in Hampshire in 2010. The fatal fire investigation report provides some additional detail: www.hantsfire.gov.uk/shirleytowers.pdf

In 2010, the Health and Safety Executive set out the requirement for fire and rescue services to prepare their personnel adequately to make decisions in environments which are dangerous and fast paced, and sometimes lacking complete or accurate information. They make clear that the quality of decision-making will reflect on how effectively the fire and rescue service has prepared that person. For more information see their 2010 publication *Striking the Balance Between Operational and Health and Safety Duties in the Fire and Rescue Service*: www.hse.gov.uk/services/fire/duties.pdf

Joseph LeDoux's 2003 book *The Synaptic Self: How Our Brains Become Who We Are* (New York: Penguin) offers a fascinating overview on how our brains – our neurons and synapses – operate to encode the essence of who we are, our personalities, our memories, and how we respond from moment to moment.

Further reading on Pavlovian conditioning can be found in the following book: Dickinson, A., (1980). *Contemporary Animal Learning Theory*. Cambridge: Cambridge University Press.

Instrumental conditioning is also known as operant behaviour. B. F. Skinner is widely regarded as the grandfather of instrumental conditioning following his 1935 publication: 'The Generic Nature of the Concepts of Stimulus and Response'. *The Journal of General Psychology*, 12, 40–65.

For a great review of the literature that relates Pavlovian-instrumental Transfer from a neurobiological perspective, see the following paper: Holmes, N. M., Marchand, A. R., and Coutureau, E. (2010). 'Pavlovian to Instrumental Transfer: A Neurobehavioural Perspective'. *Neuroscience and Behavioural Reviews*, 34, 1277–95.

Chapter 4 – *The Jigsaw*

A report for the Business Sprinkler Alliance demonstrates the economic and knock-on effects of fires in warehouses, which included a direct financial loss of £230.2 million to the affected businesses and cost nearly 1,000 job losses per year. Further information on the report 'The Financial and Economic Impact of Warehouse Fires' can be found here: www.business-sprinkler-alliance.org/wp-content/uploads/2014/01/Cebr-BSA-report.pdf

Further information on theoretical aspects of situational awareness can be found in the following paper: Endsley, M. R. (1995). 'Toward a Theory of Situation Awareness in Dynamic Systems'. *Human Factors*, 37, 32–64.

Pat Croskerry, MD, PhD is a professor in emergency medicine at Dalhousie University, Halifax, Canada. In addition to his medical training, he holds a doctorate in experimental psychology. His research is principally concerned with clinical decision-making, specifically on diagnostic error. For a great video that highlights some of the common applications and errors associated with critical thinking from the perspective of a medical environment, see the following: https://www.riskybusiness.events/talk.php?id=344

For further information about the role of situational awareness in incident command, see National Operational Guidance for Incident Command. More detail can be found here: www.ukfrs. com/foundation-knowledge/foundation-incident-command? bundle=section&id=16988&parent=16977

Chapter 5 – Trust Your Gut

The role of intuition in decision-making has been explored in a field of psychology called naturalistic decision-making. One of the founding researchers in the field is Dr Gary Klein. An overview of intuitive decision-making and its role in everyday life is given in his 2003 book, *Intuition at Work* (New York: Doubleday).

The idea that people make decisions in a rational, analytical way is reflected in many normative models of decision-making. For an overview, see the following publication: Dewey, J., (1933). *How We Think: A Restatement of the Relation of Reflective Thinking to the Educative Process.* Boston: D. C. Heath.

In this chapter I talk about one of my studies where we identified that most of the decisions made by commanders are intuitive. For more detail see the following paper: Cohen-Hatton, S. R., Butler, P. C., and Honey, R. C. (2015). 'An Investigation of Operational Decision Making in Situ: Incident Command in the UK Fire and Rescue Service'. *Human Factors,* 57(5), 793–804.

There is an enormous body of work exploring the neural basis of decision-making. For a brief snapshot that relates to the neural basis of biases and of rational (or analytical) decisions, see the following reference: De Martino, B., Kumaran, D., Seymour, B., and Dolan, R. J. (2006). 'Frames, Biases, and Rational Decision-making in the Human Brain'. *Science*, 313(5787), 684–687. For a more in-depth description of the factors that affect decision-making and the related neurobiological structures, see the following: Doya, K. (2008). 'Modulators of Decision Making'. *Nature Neuroscience*, 11(4), 410.

The research provided an evidence base for the National Operational Guidance for Incident Command, which can be found here: www.ukfrs.com/guidance/incident-command. The accompanying foundation document can be found here: www.ukfrs.com/foundation-knowledge/foundation-incident-command

The Chief Fire Officer Association (now the National Fire Chiefs Council) published a report that detailed the research and its application to the current position for command, and outlined potential future challenges. This report can be found here: www.nationalfirechiefs.org.uk/write/MediaUploads/NFCC%20Guidance%20publications/Operations/CFOA_Incident_Command_future.pdf

I refer to the Holborn fire in 2015. More information about the fire can be found in the following news report: www.bbc.co.uk/news/uk-england-london-32173689

The national policy document that fire and rescue services use is called National Operational Guidance. This document sets out the use of the Decision Control Process. More detail can be found here: www.ukfrs.com/guidance/search/command-decision-making

The Joint Emergency Services Interoperability Principles are designed to improve the way in which responding agencies work together, particularly at complex and major incidents.

The joint doctrine that sets out the principles, including the use of the decision control techniques, can be found here: www.jesip.org.uk/uploads/media/pdf/Joint%20Doctrine/ JESIP_Joint_Doctrine_Document.pdf

Chapter 6 – Judged on Shadows

In this chapter I discuss the implications of group dynamics and their powerful effects on situations and on people. For a review of the psychological theory and progress over the last hundred years, see '100 Years of Groups Research', the special issue of the journal *Group Dynamics: Theory, Research, and Practice*: Forsyth, D. R., ed. (2000). *Group Dynamics: Theory, Research, and Practice*, 4(1).

Further information on the Revised NEO Personality Inventory can be found here: www.hogrefe.co.uk/shop/neo-personality-inventory-revised-uk-edition.html, and a scholarly article discussing it can be found here: https://pdfs.semanticscholar. org/c2d3/dabde2e40e2985ddc7f211975aed820307e3.pdf

Professor Amy Edmondson, of Harvard Business School, studies the impact of psychological safety and trust on the performance of teams. For a great paper that explores these constructs, see the following: Edmondson, A., (2004). 'Psychological Safety, Trust and Learning: A Group-level Lens'. In Kramer, R., and Cook, K., eds, *Trust and Distrust in Organizations: Dilemmas and Approaches*. New York: Russell Sage Foundation, 239–272. She also gave a brilliant TEDx Talk about psychological safety within teams and the importance of speaking up: www.youtube.com/watch?v=LhoLuui9gX8

In this chapter I discuss crew resource management. For an interview with one of the founding developers of the approach, John Lauber, who gives a great overview of some of the air disasters that provided the impetus for the focus on human factors, see the following: www.vanityfair.com/news/business/2014/10/air-france-flight-447-crash

For more information on the nature versus nurture debate and what we know from research, see the following: Plomin, R. (1994). *Genetics and Experience: The Interplay Between Nature and Nurture.* Thousand Oaks: Sage Publications, Inc.

Chapter 7 – Mental Preparation

The principles by which emergency services operate at terrorist incidents are sensitive, so there is not a great deal of further reading available. However, the Kerslake report scrutinized the response by emergency services to the Manchester Arena bombing in 2017. Between pages 24 and 29, there is a more in-depth explanation of the cold, warm and hot zones, and the restrictions associated with each: www.kerslakearenareview.co.uk/media/1022/kerslake_arena_review_printed_final.pdf

For more information on the terror attacks listed in this chapter, please see: www.bbc.com/news/av/world-south-asia-14662732/2008-mumbai-attacks (Mumbai attack in 2008); www.theguardian.com/world/2010/jan/07/egypt-gunmen-kill-coptic-christmas (Nag Hammadi cathedral massacre in Egypt, 2010);www.theguardian.com/australia-news/2014/dec/20/sydney-siege-timeline-how-a-day-and-night-of-terror-unfolded-at-the-lindt-cafe (Sydney café besieged for sixteen hours in 2014);

www.bbc.co.uk/news/world-africa-33304897 (shooting on a Tunisian beach in 2015); www.bbc.co.uk/news/world-europe-30708237 (the attack at the offices of *Charlie Hebdo* in 2015); www.bbc.co.uk/news/world-europe-34827497 (the attacks on the Bataclan theatre in Paris in 2015); and www.bbc.co.uk/news/live/world-us-canada-33181651 (the attack on worshippers at the Charleston Church in 2015).

I emphasize the importance of shared situational awareness. The following paper includes a useful section on the critical nature of a common operating picture for those dealing with emergency incidents: House, A., Power, N., and Alison, L. (2014). 'A Systematic Review of the Potential Hurdles of Interoperability to the Emergency Services in Major Incidents: Recommendations for Solutions and Alternatives'. *Cognition, Technology & Work*, 16(3), 319–335. This is further described in Salas E., and Cannon-Bowers, J. A. (2001). 'The Science of Training: A Decade of Progress'. *Annual Review of Psychology*, 52, 471–499.

My TEDx Talk is available to view here: www.youtube.com/watch?v=-2TsVBLcEmE. You can decide for yourself whether the blank moments could be interpreted as pregnant pauses!

A section of the 'Future of Incident Command' report issued by the Chief Fire Officer Association (now the National Fire Chiefs Council) detailed the future challenges posed by falling experience levels and sets out a series of recommendations for fire and rescue services to consider. This report can be found here: www.nationalfirechiefs.org.uk/write/Media Uploads/NFCC%20Guidance%20publications/Operations/CFOA_Incident_Command_future.pdf

I outline some work that we did in the National Decision Trials. The results of that study were published in the following research paper: Cohen-Hatton, S. R., and Honey, R. C. (2015).

'Goal-oriented Training Affects Decision-making Processes in Virtual and Simulated Fire and Rescue Environments'. *Journal of Experimental Psychology: Applied*, 21(4), 395.

More information on the Hydra system and the associated Hydra Foundation run by Jonathan Crego can be found here: www.hydrafoundation.org

This chapter describes research carried out during the evaluation of Exercise Unified Response. The full evaluation report can be accessed here: www.london-fire.gov.uk/media/3008/lfb-evaluation-eur-report.pdf. The findings from the research are in the following publications [in press]: Wilkinson, B., Cohen-Hatton, S. R., & Honey, R. C. (2019). 'Decision-making in multi-agency groups at simulated major incident emergencies: In situ analysis of adherence to UK doctrine', *Journal of Contingencies and Crisis Management*: https://doi.org/10.1111/1468-5973.12260; and also in Wilkinson, B., Cohen-Hatton, S. R., and Honey, R. C. (2019). 'Variation in Exploration and Exploitation in Group Decision Making: Evidence from Immersive Simulations of Major Incidents' [under review].

The use of the decision controls in multi-agency environments are set out within the Joint Emergency Services Interoperability Principles. The doctrine that sets out the principles, including the use of the decision control techniques, can be found here: www.jesip.org.uk/uploads/media/pdf/Joint%20Doctrine/JESIP_Joint_Doctrine_Document.pdf

Chapter 8 – *The Price of Being Human*

I reference some statistics that relate to the prevalence of post-traumatic stress disorder (PTSD) in the general population.

Further information on these figures can be found in the Mental Health and Wellbeing in England Adult Psychiatric Morbidity Survey (2014), which can be accessed here: http://webarchive.nationalarchives.gov.uk/20180328140249/http://digital.nhs.uk/media/35660/APMS-2014-Full-Report/pdf/Mental_health_and_wellbeing_in_England_full_report

I discuss some of the findings from the Mind Blue Light Research Programme (2016). A summary of the research can be accessed here: www.mind.org.uk/media/4614222/blue-light-programme-research-summary.pdf

Further information on PTSD for emergency responders can be found here: www.mind.org.uk/news-campaigns/campaigns/bluelight/ptsd-and-trauma

I refer to the Blue Light Champions programme initiated by Mind. Further information can be found here: www.mind.org.uk/media/2416377/blp-role-description.pdf

Chapter 9 – Brains and Behaviour

I discuss the incident in which Alison Hume lost her life and the ensuing comments from the inquiry. More information can be found in a report published by Her Majesty's Fire Service Inspectorate (2012), 'A Report to Scottish Ministers: The 2008 Galston Mine Incident', Edinburgh: APS Group Scotland.

I mention the work of neuroscientist Uri Hasson and his work on the way your brain works when you are communicating with other people. For a great overview of his work in this area, see his TED Talk, 'This Is Your Brain on Communication': www.ted.com/talks/uri_hasson_this_is_your_brain_on_communication

I reference the explosion caused by the mass storage of fire-works at Marlie Farm in East Sussex in 2006 which killed two (also referenced in Chapter 3). The following report outlines the significant findings from the incident: www.esfrs.org/news/2011/july/20110726MarlieFarmSignificant FindingsReport.shtml

Two firefighters lost their lives while fighting a fire at Shirley Towers in Hampshire in 2010 (also referenced in Chapter 3). The fatal fire investigation report provides some further detail around the circumstances of the incident: www.hants-fire.gov.uk/shirleytowers.pdf

I refer to a behavioural marker scheme that we have developed for the fire and rescue service. For associated research papers, see the following: Butler, P. C., Honey, R. C., and Cohen-Hatton, S. R. (2019). 'Development of a Behavioural Marker System for Incident Command in the UK Fire and Rescue Service: THINCS'. *Cognition, Technology & Work*: https://doi.org/10.1007/s10111-019-00539-6

Chapter 10: Hindsight is Twenty-twenty

The study by Elizabeth Loftus in which she demonstrated that false memories could be implanted in the minds of children can be found here: Ceci, S. J., Loftus, E. F., Leichtman, M. D., and Bruck, M. (1994). 'The Possible Role of Source Misattributions in the Creation of False Beliefs Among Preschoolers'. *International Journal of Clinical and Experimental Hypnosis*, 42(4), 304–320. Her study implanting memories using simple adverts (in this case relating to a visit to Disneyland) can be read in full here: Braun, K. A., Ellis, R., and Loftus, E. F.

(2002). 'Make My Memory: How Advertising Can Change Our Memories of the Past'. *Psychology & Marketing*, 19(1), 1–23.

For an overview of Loftus's work, see her TED Talk, 'How Reliable Is Your Memory?': www.ted.com/talks/elizabeth_loftus_the_fiction_of_memory/discussion?en

Further information on the 10,000 volts debriefing system can be found here: www.hydrafoundation.org/hydra-methodology/10-000-volt-debriefing

I describe the Regulation 28 reports that coroners can issue to prevent future deaths. An example of a Regulation 28 letter issued to fire and rescue services can be found in the link below. This report relates to the fatality of firefighter Stephen Hunt, who was killed in a fire in Manchester in 2013, and sets out a series of recommendations for fire and rescue services to consider: www.judiciary.uk/wp-content/uploads/2016/09/Hunt-2016-0216.pdf

Acknowledgements

The journey to writing this book and to completing the research that came before it has been littered with sacrifices. There are the obvious ones, such as sleep, but these were mainly my choice. The greatest sacrifices, however, were made by other people. They did not opt in, but nonetheless made sacrifices to allow me to undertake these projects and endured them with grace, patience, love and support. And the good news? If you're reading this, then we did it.

There are three people without whom this book would never have been possible.

Gabriella, everything I do is for you; always has been and always will be. If ever there was any doubt that I'd finish writing this book, you set it in flames when you declared to your friends at the school gate, 'Mummy's going to be an author! Of an actual book!' Which was much better than your early phrases as a toddler that were something like, 'Mummy's 'speriment didn't work.' Even as a baby you were insightful beyond your years! I promise to raise you as though you breathe fire, so you can go through life burning up the doubts or challenges that are thrown your way. I love you unreservedly.

Mike, without you and your unwavering support for every crazy idea I've had since the very second we met, this would have never been more than a passing idea over a coffee and a slice of carrot cake. You have been my rock. I couldn't have got here without you. I love you and I'm forever grateful.

My father will never read this because he was taken from us far too soon. Dad, my insatiable curiosity for neuroscience arose from your illness. It is a privilege to contribute, in some small way, to some small part of the field that works tirelessly to improve the prognosis for people like you, and families like ours. I love you always.

Professor Rob Honey has been my supervisor, mentor and collaborator. He has guided and never dogmatized, and always inspired me to learn. Thank you for enabling me to undertake this research and for the incredible support I've received from you and Cardiff University. You saw past my circumstances and knew all along that it doesn't matter where someone starts, but where they go. You've been amazing.

I have an enormous thank you to say to my editor, Lizzy Goudsmit. Your endless patience and advice took the seedlings of ideas and helped me to refine them into something much more meaningful that I ever would have produced alone. Thank you for your help, guidance and support. And to my agent, Kirsty McLachlan. The one brief email exchange that brightened up an otherwise very grey day was the precursor to all of this. Your guidance through a brand-new industry has been invaluable. Rebecca Wright, your fresh eyes and editing expertise was the icing on the cake. Thank you also to the most wonderful publicist, Becky Short; marketer, Ella Horne; and cover designer, Beci Kelly. Ladies, I salute you. With heartfelt thanks.

There are several other people who have been absolutely

instrumental. Simon Pilling, who was the National Command and Control Lead when I first started my research, saw the benefit in trying something new. He provided an environment in which innovation could flourish in an otherwise highly traditional sector. Jamie Courtney has picked up where Simon left off and continues to encourage us to push new boundaries. Roy Wilsher – now Chair of the NFCC – is an ongoing support and his leadership has been unwavering. He has allowed us to disseminate our findings across the globe and to improve the safety of emergency workers worldwide. The other members of the ICS team – John Baines, Phil Butler, Kevin Hepple, Adrian Brown and Simon Barry – have been amazing collaborators, along with Jim Clarke, Kirk Cornwall, Byron Wilkinson, Keith Smith and the Fire Service College. Thank you to every single incident commander who participated in any one of our studies. Your willingness to step outside the everyday was invaluable.

I need to express special thanks to Ron Dobson, who brought me into London – a city that I've fallen in love with and which has become home for both me and my family. It's a city that has given me the most incredible experiences. Also, my gratitude goes to Huw Jakeway for his unwavering support for the research since its conception – from the point when I was juggling being an officer and doing a PhD – right through to the present day. A heartfelt thank you also goes to Hampshire Fire and Rescue Service; in particular, Andy Bowers. The support from everyone – crews, officers and firefighters alike – has been unwavering, and the proactive and innovative approach you take towards training made the crucial live-burn phase possible. To Neil Odin, I'd like to say thanks for lending them to me!

I have had some brilliant mentors throughout my time in the service, some of whom guided me through choppy waters and others who have proved to be great friends. Some I have already mentioned, such as my dear Jonathan Crego and Simon Pilling, but I also want to thank Dave Etheridge and Brian Fraser. Gentlemen, I am eternally grateful.

I also want to thank the Biotechnology and Bioscience Research Council (BBSRC), who recognized the impact of our work, which they part funded, by naming us as Innovators of the Year 2018, and the Economic and Social Research Council (ESRC), who are funding the research we are carrying out into behavioural markers and strategic decision-making.

A final thank you must go out to two people whose support made writing this book a reality. Dany Cotton and Steve Apter. Thanks for backing me on this journey from the minute I knocked on the door and said, 'Um . . . I have an idea . . .' Your support and advice meant the world. And last, but by no means least, Glenn Sebright, thanks for taking the time and trouble to impart your sage advice!

About the Author

Dr Sabrina Cohen-Hatton has been a firefighter since she was eighteen years old and is now one of the most senior female firefighters in the UK.

After leaving home at fifteen and school at sixteen, she joined the fire and rescue service in Wales. While climbing the ranks, she studied at the Open University, Middlesex University and then at Cardiff University, eventually completing a PhD in Psychology. She has won awards for her research into incident command in the emergency services and has presented her findings across the world. She was recently conferred as an Honorary Fellow at Cardiff University.

She lives with husband, daughter and dog (Jimmy Chew) in London.